How to Pass

SECOND EDITION

NATIONAL 5

Geography

Dr Bill Dick

Consultant Editor: Sheena Williamson

D1340670

HODDER
GIBSON

AN HACHETTE UK COMPANY

Dedication: For all my family. (BD)

The Publishers would like to thank the following for permission to reproduce copyright material:

Photo credits

p.45 © Kenny Williamson/Alamy Stock Photo; **p.46** (left) © Doug Houghton SCO/Alamy Stock Photo; (right) © cornfield/ Shutterstock; **p.87** (left) via realbenidorm.net; (right) © valery121283/stock.adobe.com

Acknowledgements

The table on pages vi–vii is from the National 5 Geography Course Assessment specification are reproduced with permission © Scottish Qualifications Authority.

Every effort has been made to trace all copyright holders, but if any have been inadvertently overlooked the Publishers will be pleased to make the necessary arrangements at the first opportunity.

Although every effort has been made to ensure that website addresses are correct at time of going to press, Hodder Gibson cannot be held responsible for the content of any website mentioned in this book. It is sometimes possible to find a relocated web page by typing in the address of the home page for a website in the URL window of your browser.

Hachette UK's policy is to use papers that are natural, renewable and recyclable products and made from wood grown in sustainable forests. The logging and manufacturing processes are expected to conform to the environmental regulations of the country of origin.

Orders: please contact Bookpoint Ltd, 130 Park Drive, Milton Park, Abingdon, Oxon OX14 4SE. Telephone: (44) 01235 827720. Fax: (44) 01235 400454. Lines are open 9.00–5.00, Monday to Saturday, with a 24-hour message answering service. Visit our website at www.hoddereducation.co.uk. Hodder Gibson can be contacted directly at hoddergibson@hodder.co.uk

Cover photo © Milosh Kojadinovich/123RF
Illustrations by Tony Wilkins Design, Jeff Edwards and Aptara, Inc.
Typeset in 13/15 Cronos Pro by Aptara, Inc.
Printed in Spain
A catalogue record for this title is available from the British Library
ISBN: 978 1 510 42091 5

MIX
Paper from
responsible sources
FSC™ C104740

Contents

Introduction

The exam

The overall course assessment is worth 100 marks. This is made up of two parts: a question paper worth 80 marks and an assignment worth 20 marks. The question paper makes up 80 per cent of the total marks, with the exam being completed in 2 hours and 20 minutes. The assignment is worth 20 per cent. It is completed throughout the year and submitted to SQA around the end of March. The marks for the two parts are added together to give you a final award.

The question paper

The purpose of the question paper is to allow you to demonstrate the skills you have acquired and to reveal the knowledge and understanding you have gained from the topics studied throughout the course. The question paper will give you the chance to show your ability in describing, explaining, matching and evaluating a broad range of geographical information, as well as using a variety of maps and demonstrating proficiency in Ordnance Survey (OS) skills. Candidates will complete this question paper in 2 hours and 20 minutes. Questions will be asked on a local, regional and global scale. The question paper has three sections.

Section 1: Physical environments

This section is worth 30 marks. Candidates will answer a mixture of limited/ extended response questions by using the knowledge, understanding and skills learnt throughout the course. In this section there is a choice. Candidates should answer **either** Q1: glaciation/coasts **or** Q2: rivers/ limestone. This will be dependent on the topics taught at your school. Some topics you could be asked to answer questions on include **Weather**, **Landscape formations** within Scotland and/or the UK, **Land use management** conflicts and solutions. In this section you may also be examined on Ordnance Survey skills using a map.

Section 2: Human environments

This section is worth 30 marks. Candidates will answer a mixture of limited/ extended response questions by using the knowledge, understanding and skills learnt throughout the course. Candidates should answer all questions in this section. Questions in this section are drawn from both the developed and the developing world. Some topics you could be asked questions on include **Population** (development indicators, population distribution, factors affecting birth rates and death rates), **Urban** (land use characteristics in cities in the developed world, recent developments in developed world cities, strategies to improve shanty towns) and **Rural** (changes in rural landscapes in both the developed and developing world). In this section you may also be examined on Ordnance Survey skills using a map.

Section 3: Global issues

This section is worth 20 marks, made up of two 10-mark questions. Candidates will answer a mixture of limited/extended response questions by using the knowledge, understanding and skills learnt throughout the course. In this section there is a choice of questions. Candidates should answer **two** questions from a choice of six. Your choice will be dependent on the topics taught at your school. The topics you will have to choose from are: **Climate change, Natural regions, Environmental hazards, Trade and globalisation, Tourism** and **Health**.

Types of questions

The main types of questions used in the paper are: Describe, Explain, Give reasons, Match, Give advantages and/or disadvantages and Give map evidence.

The assignment

The assignment will allow you to demonstrate your skills and knowledge of a geographical topic or issue of your choice. The assignment concentrates more on the examination of skills than the exam paper. It is worth 20 marks out of a total of 100 marks. Your assignment can be field work based where you go out and collect first-hand information or a classroom-based assignment where you use books, internet, articles etc. While you can collect information for your assignment in a group, you must process and analyse your information on your own.

Research stage

There are four main steps in the research stage of your assignment:
1 Choose a topic
2 Collect information
3 Process the information gathered
4 Describe and explain your findings

Evidence stage

When you have completed the research stage of you assignment, you will be required to write it up under exam conditions. You will have 1 hour to write up your findings. You should have produced two single sides of A4 or a single side of A3 paper, containing your processed information from your research, to help you write up your assignment. This will be written up at an appropriate time and submitted by a date set by SQA, usually before the end of March.

Some examples of assignment topics include:
● comparison of the upper and lower courses of a river
● urban land use study
● comparison of environmental quality
● impact of an earthquake or tropical storm
● farm study.

What is covered in the National 5 Geography course?

The following information is taken from the National 5 course specification provided by the Scottish Qualifications Authority (SQA).

Section 1: Physical environments
Weather
Within the context of the United Kingdom: ● the effect of latitude, relief, aspect and distance from sea on local weather conditions ● the characteristics of the five main air masses affecting the UK ● the characteristics of weather associated with depressions and anticyclones.
Landscape types
Within the context of **two** landscape types, selected from either: ● glaciated uplands and coastal landscapes, or ● upland limestone, and rivers and their valleys. The identification and formation of the following landscape features (from **two** landscape types): ● glaciated uplands – corrie, truncated spur, pyramidal peak, arête, u-shaped valley ● upland limestone – limestone pavements, potholes/swallow holes, caverns, stalactites and stalagmites, intermittent drainage ● coastal landscapes – cliffs, caves and arches, stacks, headlands and bays, spits and sand bars ● rivers and their valleys – v-shaped valleys, waterfalls, meander, oxbow lake, levee. Land uses appropriate to the **two** landscape types studied. The land uses should be chosen from: ● farming ● forestry ● industry ● recreation and tourism ● water storage and supply ● renewable energy. In relation to **one** landscape type studied, candidates should be able to describe and explain: ● the conflicts which can arise between land uses within this landscape ● the solutions adopted to deal with the identified land use conflicts.

Section 2: Human environments
In the context of developed and developing countries: ● social and economic indicators ● physical and human factors influencing global population distribution ● factors affecting birth and death rates. In the context of urban areas: ● characteristics of land use zones in cities in the developed world ● recent developments in the CBD, inner city, rural/urban fringe in developed world cities ● recent developments which deal with issues in shanty towns in developing world cities. In the context of rural areas: ● changes in the rural landscape in developed countries, related to modern developments in farming such as: diversification, impact of new technology, organic farming, GM, current government policy ● changes in the rural landscape in developing countries related to modern developments in farming such as: GM, impact of new technology, biofuels.

Section 3: Global issues

Candidates should study **two** global issues from the following:

Climate change

- features of climate change
- causes – physical and human
- effects – local and global
- management strategies to minimise impact/effects.

Natural regions

- description of tundra and equatorial tropical forest climates and their ecosystems
- use and misuse of these environments by people
- effects of land degradation on people and the environment
- management strategies to minimise impact/effects.

Environmental hazards

- the main features of earthquakes, volcanoes and tropical storms
- causes of each hazard
- impact of each hazard on people and the landscape
- management – methods of prediction and planning; and strategies adopted in response to environmental hazards.

Trade and globalisation

- world trade patterns
- cause of inequalities in trade
- impact of world trade patterns on people and the environment
- strategies to reduce inequalities – trade alliances, fair trade, sustainable practices.

Tourism

- features of mass tourism and ecotourism
- causes of/reasons for mass tourism and ecotourism
- impact of mass tourism and ecotourism on people and the environment
- strategies adopted to manage tourism.

Health

- distribution of a range of world diseases
- causes, effects and strategies adopted to manage:
  - HIV/AIDS in developed and developing countries
  - one disease prevalent in a developed country (choose from: heart disease, cancer, asthma)
  - one disease prevalent in a developing country (choose from: malaria, cholera, kwashiorkor, pneumonia).

Geographical skills

The following skills will be assessed in contexts drawn from across the course:

Mapping skills including the use of Ordnance Survey maps:

- grid references (4/6 figure)
- identification and location of physical and human features and patterns
- measure distance using scale
- interpret relief and contour patterns
- use maps in association with photographs, field sketches, cross sections/transects.

Extracting, interpreting and presenting numerical and graphical information which may be:

- graphs
- tables
- diagrams
- maps.

In addition to the above, the assignment will also cover research skills, including the fieldwork skills of:

- gathering
- processing
- interpreting
geographical data.

Maps and mapping skills

Some marks available in the question paper will be based on an Ordnance Survey map at a scale of either 1:50 000 or 1:25 000. The questions asked could be a mixture of knowledge and understanding and skills, particularly those relating to physical landscape, land use patterns and land use conflicts.

These questions are designed to test your mapping skills, which you will have been taught throughout your courses at both National 4 and National 5. Note that it is only at National 5 that these skills will be assessed in an external exam.

There are certain basic skills that you should have and that you should practise. You should be able to:

- give four- and six-figure grid references
- recognise symbols, although the OS maps will have a key
- recognise contour patterns in describing physical landscapes
- relate land use to the physical landscape
- understand and use cross-sections
- identify and locate physical and human features
- measure distance using scale
- interpret relief and contour patterns
- use maps in association with photographs, field sketches and cross-sections/transects.

Mapping skills in internal and external exams

Questions in internal and external exams will ask you to give map evidence in support of your answers. Do this by using appropriate grid references to locate relevant features and refer to features on the map so as to show that you are actually using the map in your answer. It is advisable to give a number of grid references in your answer.

There may be a variety of questions that include requests for descriptions of landforms, features and patterns, identification of certain physical and human features, or your opinion based on map evidence on how various features interact with each other. When answering these questions you are combining your knowledge of these topics with mapping skills. The length of your answer and the detail given should be related to the number of marks available. By answering the question with examples from the map and using grid references, you should score high marks.

If you get to the end of an exam and find that you have some time left, go back to the map question and go over your answers, adding additional detail and evidence to any answers that may be a little short.

You may be asked to provide labelled diagrams, for example if asked to explain the formation of certain physical landforms that you have identified on the map.

Gathering skills

The skills that you should be able to demonstrate in the assignment include:

- extracting information from maps
- field sketching
- measuring
- recording information on a map (land use, location, distribution)
- observing and recording (traffic/pedestrian flows, weather, environmental quality)
- compiling and using questionnaires and interviews.

There may be other skills, such as taking photographs, that could be appropriate in some situations.

Processing techniques

The techniques that you should know about include:

- classifying, tabulating and matrixing information
- drawing graphs (bar, multiple bar, line, pie and scatter)
- drawing maps (land use, location and distribution)
- drawing cross-sections/transects
- annotating or labelling maps, graphs, field sketches and photographs.

As with gathering techniques, there may be other techniques that you could mention, such as drawing pictograms, which might apply in certain situations and for which you would be given credit.

Some revision tips

- Make sure you have a copy of the examination timetable and use this to work out a schedule for studying with a programme, which includes what sections of the syllabus you intend to study and when.
- Try to avoid leaving your studying to a day or two before the exam. Also try to avoid cramming your studies on the night before the examination, especially staying up late to study.
- Organise your notes into checklists and revision cards. One useful technique when revising is to use summary note cards on individual topics.
- Make use of questions from past papers to test your knowledge of the topics and your exam technique. Go over your answers and give yourself a mark for every correct point you make when comparing your answer with your notes. If you work with a classmate, try to mark each other's practice answers.
- Practise your writing skills. Try to ensure that your answers are clearly worded. Try to develop the points which you make in your answer.
- Practise drawing diagrams which may be included in your answers, for example, corries or pyramidal peaks.

Some tips for the exam

- Make sure you know the examination timetable, noting the dates and times of your examinations. Give yourself plenty of time by arriving early for the examination, well equipped with pens, pencils, rubbers and so on.

- Make sure that you have read the instructions carefully and that you have avoided needless errors such as answering the wrong sections.
- Read the questions very carefully. If the question asks you to 'describe' make sure that this is what you do. If you are asked to 'explain' you must use phrases such as 'due to', 'this happens because', 'this is a result of'. If you describe rather than explain, you will lose most of the marks for that question.
- Use the number of marks as a guide to the length of your answer. For example, if a question is worth 4 marks then you should make four points. If it is worth 6 marks, then you should make six points.
- Try to include examples in your answer wherever possible. If asked for diagrams, draw clear, labelled diagrams.
- Try to avoid lists as, in general, only 1 mark would be awarded for a list answer.
- If you are asked to explain the suitability of a development, such as an industrial estate or a housing area from an OS map, you can give both good and bad points.
- If you have any time left at the end, use this time productively by going back over your answers and perhaps adding additional parts to your answer. This is especially helpful in Ordnance Survey map based questions.
- One technique which you might find helpful is to think of all possible points to include before writing your answer. You can write these down in a list at the start of your answer. Then, when you start writing your answer, you can double check with your list to ensure that you have put as much into your answer as you can. It avoids coming out of the exam and being annoyed that you forgot to mention an important point.

Common exam errors

Common errors in the external exam include the following:

- **Lack of sufficient detail** – This often occurs in higher mark questions. Many candidates fail to provide sufficient detail in answers by omitting reference to specific examples, or failing to elaborate or develop points made in their answer. As noted earlier a good guide to the amount of detail required is the number of marks given for the question.
- **Listing** – If you give a simple list of points rather than fuller statements in your answer you will automatically lose marks; for example in a 4-mark question you might obtain only 1 mark for a list.
- **Bullet points** – The same rule applies to a simple list of bullet points.
- **Irrelevant answers** – You must read the question instructions carefully so as to avoid giving answers which are irrelevant to the question. For example, if asked to 'explain' and you 'describe' then you will lose marks. If asked for a named example and you do not provide one, you will lose marks.
- **Repetition** – You should be careful not to repeat points already made in your answer. These will not gain any further marks. You may feel that you have written a long answer, but it may contain the same basic information repeated again and again.

Section 1: Physical environments

Chapter 1.1
Weather

Weather will be examined within the context of the United Kingdom.

The topics outlined in the Introduction are built on a more basic knowledge of weather, which you will acquire during the National 4 course. You will not be examined on this basic knowledge in the National 5 examination but you will be expected to refer to National 4 course content in some answers.

This basic knowledge includes:
- the main elements of weather
- how these elements are measured
- the instruments used to measure various elements
- the appropriate locations for **weather stations**
- how to record weather information using recognised symbols.

The following section begins with a review of this National 4 knowledge.

Elements of weather

Key points

* Elements of weather can be identified, observed, measured, recorded and classified.
* As a result, dynamic patterns of weather can be identified and used for forecasting.
* As well as knowledge of weather elements, you should also have awareness of the methods and instruments used to measure them and all the units in which they are recorded.

- Weather elements are the various features that contribute to weather.
- Weather itself is the day-to-day conditions of the atmosphere.
- There is a wide variety of ways in which weather information can be recorded. The basic method is to use instruments including **thermometers**, **barometers**, **wind vanes**, **anemometers** and **rain gauges** set up at a weather station. Other methods include radiosonde balloons, aircraft, ships and satellites.

- Information from these sources is gathered every day and sent to meteorological offices. Here the data is recorded, analysed and transferred onto weather maps known as **synoptic charts**.

Weather symbols

Information for individual weather stations can be shown using a variety of symbols drawn around a circle that denotes the weather station. At National 4 and National 5 you should know these symbols and be able to describe weather conditions from given weather station circles such as that shown in Figure 1.1.

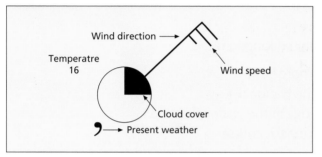

Symbol	Precipitation	Symbol	Cloud cover	Symbol	Wind speed
,	Drizzle	○	Clear sky	◎	CALM
▽	Shower	◑	One **okta**	○—	1–2 knots
•	Rain	◕	Two oktas	○—⌐	5 knots
*	Snow	◔	Three oktas	○—⊤	10 knots
△	Hail	◑	Four oktas	○—⊤⊤	15 knots
⚡	Thunder-storm	◕	Five oktas	○—⊤⊤⊤	20 knots
••	Heavy rain	◕	Six oktas	○—▼	50 knots or more
•••	Very heavy rain	◕	Seven oktas		
•*	Sleet	●	Eight oktas		
*▽	Snow shower	⊗	Sky obscured		
—	Mist				
≡	Fog				

Figure 1.1 Weather station circle and symbols

Factors affecting Britain's weather

Various factors have major impacts on the weather affecting Britain throughout the year. You should know these factors for National 5 assessments. They include:

Latitude

As you will know, latitude measures distance from the equator to the poles. The latitude at the equator is zero degrees, while at the North and South Poles it is 90 degrees.

The higher the latitude of a country or region, the colder the weather generally tends to be during all four seasons. Therefore Britain, which lies between latitudes 50 and 60 degrees, has much colder weather than areas further south nearer the equator.

Distance from the sea

Countries and areas that lie further from the sea tend to have drier climates. Britain, which is an island surrounded by sea, has a fairly wet climate.

Relief

The height and shape of the land can impact on the weather in an area. This happens in various ways. For example, **temperature** falls with altitude by 1 degree for every 300 metres. Therefore, temperatures in mountainous areas of Britain are generally colder than those in lowland areas.

Higher land also presents boundaries to rain-bearing winds from the west, forcing the air to rise. As it rises the temperature of the air falls, causing the wind to lose its moisture in the form of **precipitation** (rain). As a result, western parts of Britain tend to be wetter than eastern parts. This is known as a rain shadow effect.

Aspect

Aspect refers to the direction of slopes. Because of the way in which the Sun's rays strike the land in Britain, areas with south-facing slopes will receive more sunshine than north-facing areas. This obviously affects many different **land uses** in both urban and rural areas. The effect of aspect is often taken into account when choosing sites for houses and different uses for farmland.

Air masses

- Britain has often been described as a battleground so far as its weather is concerned. This is because its weather is subject to a wide range of influences, and throughout the year it is the meeting point for large areas of air coming from different parts of the world.
- The weather is affected by systems of high and low pressure.
- These systems result from the movement of large bodies of air called air masses, which originate in different parts of the world. These masses may originate in warm, cold, wet or dry regions and then travel towards the British Isles. Each of these masses of air has different characteristics in relation to pressure, moisture, temperature and so on, based on where it comes from.
- Often cold and warm air masses collide to create areas of low pressure that flow across Britain from west to east.
- Understanding these air masses helps us to understand weather systems and to predict, forecast and explain the forces that shape our weather.

● Figure 1.2 shows the main air masses that affect Britain's weather.

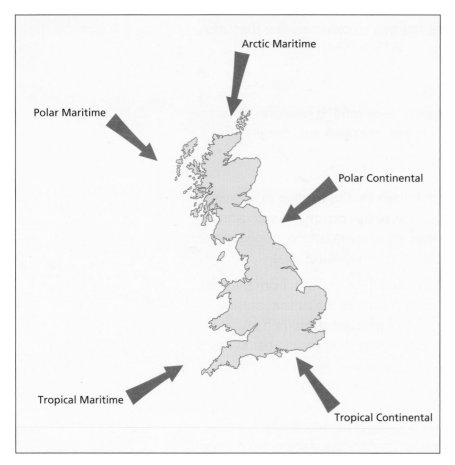

Figure 1.2 Air masses

Weather systems and weather maps

● Weather maps (synoptic charts) are constructed with lines that join up places with the same pressure. These lines are called **isobars**.
● The pattern of the isobars indicates the type of pressure system that is present. You should understand these patterns and know the kind of weather associated with them.
● Questions in the National 5 examination may ask you to explain these weather features and weather changes.

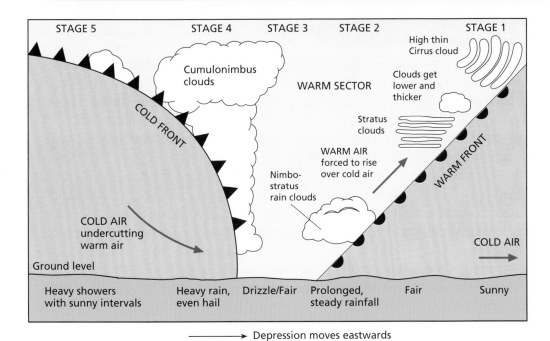

Figure 1.3 A cross-section through a low pressure system (depression)

Patterns with isobars that are close together are **low pressure** areas or **depressions** (see Figure 1.3). In order to explain weather changes caused by a depression, you need to understand the different parts of low pressure systems and the types of weather linked to them. These include:

- **warm fronts** (increasing winds, cloud cover, drizzly rain and slight rise in temperature)
- **cold fronts**, which lag behind the warm fronts (heavier winds, large clouds, heavy rain and lower temperatures)
- **occluded fronts**, found near the centre of the low pressure area (mixed conditions of rain, wind and cloud cover)
- the **warm sector**, found between the warm and cold fronts (conditions tend to settle, skies are dull and cloudy, temperatures rise slightly, there is little rain).

Where the isobars are widely spaced, this indicates **high pressure** systems or **anticyclones**.

- Anticyclones are areas of high pressure which bring calm, bright, sunny conditions. However, temperatures will depend on the time of year.
- In winter, anticyclones can bring very low temperatures in Britain. These may be below freezing, especially during the night, causing extreme frosty and occasionally foggy conditions.
- In summer, high pressure systems tend to bring periods of very warm and sunny weather.

Key words and associated terms

Air pressure: The pressure put on the Earth's surface by the atmosphere. It can vary from **low** to **high pressure**. When masses of cold air from the north meet warm air from southern latitudes, the heavier, denser cold air lifts the lighter warm air and creates an area of low pressure. Since many of these air masses form over the sea or ocean areas, they bring unsettled, wet and windy conditions as they move from west to east across the British Isles.

Anemometer: Instrument that measures wind speeds. It consists of rotating cups and a speedometer. It may be situated high up on top of a long pole or used in a smaller hand-held version.

Anticyclone: An area of high pressure that can remain in place for fairly long periods of time.

Barometer: Instrument that measures air pressure in millibars. It can be attached to a drum that has graph paper and a pen attached, to record or trace changes in air pressure over a specific period such as a week.

Cold front: The boundary between cold air and warmer air within a depression. It is usually associated with stormy and very wet conditions.

Depression: Another name for a low pressure system.

Isobar: A line on a weather map that joins places with the same pressure. It is rather similar to a contour line on an Ordnance Survey map.

Land use: This refers to how humans make use of the physical landscape, for example forestry, farming, industry or settlement.

Maximum thermometer: Instrument that allows readings to be taken of the maximum temperature reached on any given day.

Minimum thermometer: Instrument, filled with alcohol rather than mercury, that allows readings to be taken of the lowest temperature reached on any given day. By adding the maximum and minimum temperatures together and dividing by two we get the average daily temperature.

Occluded front: This occurs where the cold front overtakes the warm front, usually near the centre of a depression.

Okta: The sky is divided into eighths when describing the amount of cloud cover. One eighth is called an okta.

Precipitation: This refers to all forms of moisture in the atmosphere, for example rain, snow, sleet and hail.

Rain gauge: The instrument used to collect and measure rainfall during a 24-hour period.

Stevenson screen: A white wooden box with slatted sides in which weather instruments are placed.

Synoptic chart: A map of an area showing isobars, fronts and pressure systems from which weather forecasts can be made.

Temperature: The amount of heat in the atmosphere.

Warm front: The boundary between the edge of the warm air in a depression and the air it is replacing. This is usually associated with persistent drizzle and increasing wind and cloud cover.

Warm sector: An area of warm air between the warm and cold front in a depression.

Weather station: A site where instruments are set up to measure the various elements of the weather.

Weather station circle: A drawn circle that is surrounded by various symbols representing the weather.

Wind vane: the instrument used to measure wind speed and direction

Physical landscapes

Physical landscapes are the products of natural processes and are always changing.

Key points

* For the external exam at National 5, within the context of two different landscape types, you are expected to be able to recognise, describe and explain the formation of the main features of these landscapes.
* The two landscape types should be selected from 'glaciated uplands', 'coastal landscapes', 'upland limestone' and 'rivers and their valleys' within the British Isles.
* Make sure you fully understand what is meant by the term 'natural processes'.

Natural processes include the effects of physical and chemical **weathering** on the landscape and the impact of **erosion** and deposition processes that contribute to change in the landscape.

Physical weathering refers to the impact of elements such as:
- the weather: temperature changes, rainfall, wind, and so on
- water flowing across the land in the form of rivers and streams
- glaciation, where ice sheets moving over the land during ice ages changed the landscape.

Landscape type 1: Glaciated uplands

Key point

You should be able to recognise features of this landscape type from photographs, sketches, diagrams and Ordnance Survey maps.

Introduction

Over the last 2.5 million years, the British Isles have been covered at different times with large sheets of ice called **glaciers**. These periods are called glaciations and there may have been up to 20 different glaciations during the period known as the Ice Age. The last ice age ended about 10,000 years ago. The ice advanced southwards as the climate became colder and retreated again as the temperature increased. These advances and retreats are termed glacial and interglacial periods, respectively.

As the ice sheets moved, they changed the underlying landscape over which they passed. Two main things happened. Firstly, the ice sheets eroded (wore away) the landscape, carving out new features. Secondly, the glaciers transported the eroded material and deposited it in other parts of the country.

You should be able to identify the main features created by erosion in parts of Britain that have been subjected to glaciation. These include **pyramidal peaks**, **corries**, **u-shaped valleys**, **truncated spurs** and **arêtes**. You need to be able to identify these features on a diagram such as that shown in Figure 1.4 and on an Ordnance Survey map.

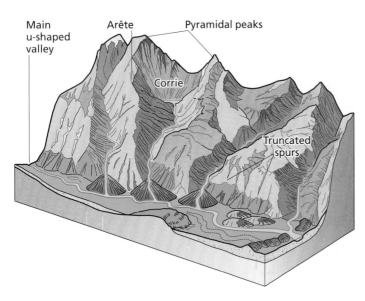

Figure 1.4 Features of glaciated upland landscapes

Upland glaciated features on a diagram or map

On a landscape diagram or an Ordnance Survey map (either with scale 1 : 50 000 or 1 : 25 000), the features you should be able to recognise from the contour patterns are those formed by erosion. You may be asked to identify the features and then describe how they were formed.

If the Ordnance Survey map is based on one of the upland glaciated areas of Britain, such as Snowdonia, the Cairngorms or the Lake District, the question may offer a choice of features. It may ask you to first identify them, and then to select one and give a detailed explanation of its formation. This kind of question could also be based on a block diagram such as that shown in Figure 1.4.

A guide to the amount of detail required is the number of marks allotted to the question. There are normally 4 marks at National 5, with 1 mark awarded for each correct statement.

You may also be asked to provide appropriate diagrams in your answer. It is quite possible to gain full marks for a well annotated (labelled) series of diagrams. Your class notes and textbooks will already have provided you with detailed descriptions and explanations of the features mentioned above. Summaries of how they were formed can be found on the next page.

Formation processes

- The general processes involved in the formation of glaciated uplands include: freeze-thaw, **plucking** and **abrasion**.
- Glacial erosion occurs through freeze-thaw, plucking and abrasion.
- Freeze-thaw is the process by which water enters the joints, cracks and hollows in rock. When the temperature reaches freezing point the water inside the cracks freezes, expands and causes the cracks to widen. When the temperature rises, the water thaws and contracts. This eventually causes rocks to break up.
- Abrasion has a sand-papering effect as the ice moves across the land. It produces smoothed surfaces.
- Plucking is where pieces of rock are torn away from the land. This happens because there is a thin film of meltwater between the glacier and the ground over which it flows. This film of water freezes and, as the glacier moves, fragments of rock are ripped or plucked from the ground. Plucking produces jagged features.
- The rate of flow of the glacier depends on the type of rock over which it flows, the amount of ice in the glacier and the slope of the land.
- As glaciers move, they erode and deposit material at their margins (the areas at their fronts and sides).
- Snow and ice are lost from the glacier through melting. This process of melting in glaciers is called ablation.
- Meltwater flowing from the glaciers further erodes and deposits material through what are known as fluvioglacial processes.

How were individual features formed?

1 Corries

- Corries are steep-sided hollows in the sides of mountains where snow accumulated and gradually compacted into ice.

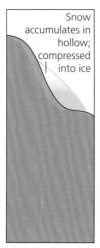
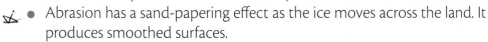
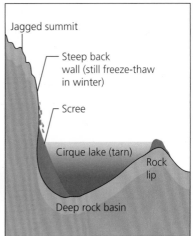

(a) Beginning of ice age

(b) During the ice age

(c) After ice age

Figure 1.5 Formation of a corrie

- The rotational movement of ice in the hollow caused considerable erosion both on the floor and the sides of the depression.
- The erosion on the floor was caused by abrasion. The floor became concave in shape and the edge took on a ridge-shaped appearance.
- At the sides of the corrie, plucking of rocks took place as the ice moved forward, and the back wall of the depression became very steep.
- As the corrie filled up with ice, eventually it could not contain any more and some moved down the slope to a lower level.
- Freeze-thaw causes fragments of rock to break off bits, sharpening the back wall of the corrie.
- Sometimes melting ice filled the corrie to form a corrie-loch or tarn.

2 Arêtes

- Corries often developed on adjacent sides of a mountain.
- When they were fully formed these corries were separated by a knife-shaped ridge termed an arête.
- The arête was then further sharpened by frost action.

3 Pyramidal peaks

- If corries developed on all sides of a mountain, the arêtes formed a jagged peak at the top.
- This feature is called a pyramidal peak. These are also further sharpened by frost action.

4 U-shaped valleys

- As a glacier moved downhill through a valley, the shape of the valley was transformed.
- A material called boulder clay was deposited on the floor of the valley.
- As the ice melted and retreated, the valley was left with very steep sides and a wide, flat floor.

Figure 1.6 Glaciated upland on an Ordnance Survey map

5 Truncated spurs

- A spur is the bottom part of a slope which juts out into the main valley. As the ice cuts through the original valley, the original spurs are removed by the ice.
- The feature that remains once the ice melts is called a truncated spur.

Figure 1.6 shows the contour pattern of a glaciated upland area.

Example

Explain the processes involved in the formation of a corrie.

You may use diagrams in your answer.

4 marks

Sample answer

Answer: Corrie

Ice and snow gather in a hollow and as this increases the bottom layers are compressed by the weight and begin to erode the hollow (✓). When the ice and snow finally melt a steep back wall is left and the original hollow is much deeper. As there is nowhere for the water to drain away to, a loch called a tarn is formed in the corrie (✓).

Comments and marks

This answer is quite basic but it contains some detail, especially about the ice and snow being compressed and eroding the hollow and melting ice being left to form a corrie-loch or tarn. The next sentence is a description and does not explain the processes involved, so no marks. A better answer might have referred to 'plucking and abrasion' and the rotational movement of the ice in the hollow helping to create the back wall and a 'lip' at the front edge of the corrie. Lack of explanation and lack of detail about the processes means this answer only achieves **2 out of 4 marks**.

Landscape type 2: Coastal landscapes

Key point !

You should be able to recognise, describe and explain features of landscapes of coastal erosion and deposition.

For examination purposes these features include: cliffs, caves, arches, stacks, headlands and bays, beaches, spits and sand bars.

Figure 1.7 shows a typical coastal landscape created by coastal erosion, consisting of headlands, caves, arches and stacks. Figure 1.8 shows these features on an Ordnance Survey map.

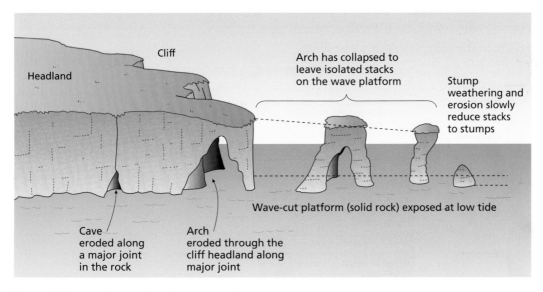

Figure 1.7 A typical coastal landscape created by coastal erosion

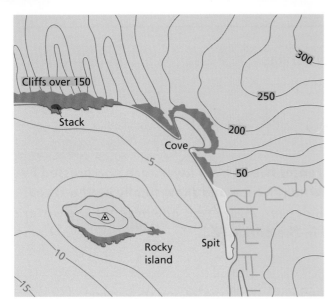

Figure 1.8 Coastal features on an Ordnance Survey map

Formation of coastal features by erosion
Cliffs

Cliffs are formed by wave action undercutting land that meets the sea. This occurs at about high tide level. A notch is cut and as the land recedes, the cliff base is deepened by wave erosion. At the same time the cliff face is continually attacked by weathering processes and mass wasting such as slumping occurs, causing the cliff face to become less steep.

When high, steep waves break at the bottom of a cliff, the cliff is undercut forming a feature called a 'wave-cut notch'. Continual undercutting causes the cliff to eventually collapse. As this process is repeated, the cliff retreats leaving a gently sloping wave-cut platform.

Headlands and bays

When resistant rocks alternate with less resistant rocks along a coast under wave attack, the resistant rocks form headlands while the less resistant rock is worn away to form bays. These can also develop in a single rock structure, for example limestone, which has lines of weakness such as joints or faults. Although the headlands gradually become more vulnerable to erosion, nevertheless they protect the adjacent bays from the effects of destructive waves.

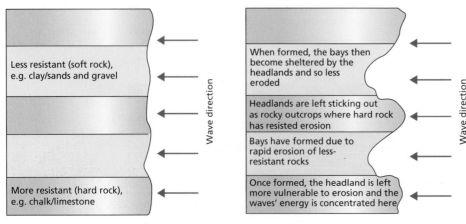

Figure 1.9 Formation of headland and bay

Caves, arches and stacks

- Caves are formed when waves attack cliffs with resistant rock along lines of weakness such as faults and joints.
- The waves undercut part of the cliff and can cut right through the cave to form an arch.
- Continual erosion causes the arch to widen. Eventually the roof of the arch collapses to leave a piece of rock left standing. This is called a stack.

Formation of coastal features by deposition ✔

Beaches

- Beaches are formed from material that is deposited by the sea.
- They form a buffer zone between the coast and the sea.
- The beach consists of loose material and its shape can be changed by factors such as wave energy, the steepness of its gradient and the seasons of the year.
- There are usually two types of beach profile: relatively wide and flat or steep and narrow.
- Wave action can be constructive or destructive and beaches can therefore be built up or destroyed by waves.
- The type of particle deposited, either sand or shingle, can also affect the shape and width of a beach.

Spits   *get diagram*

A **spit** is a long narrow ridge of sand which projects into the sea from the coastline. Material is transported along the coastline by long-shore drift. Spits form in shallow, sheltered water when there is a change in direction of the coastline. Deposition happens resulting in the accumulation of sand and shingle. The largest material is dropped first as the waves have less energy. Finer material is then deposited which builds up the spit. As the spit grows outwards, a change in wind direction can result in the spit changing direction, forming a curved end.

Sand bars

Sometimes a spit grows the whole way across a bay. A sand bar develops parallel to the shore and the waves and wind move it towards the shore until it joins the mainland. Behind the bar a lagoon is formed where water is trapped and eventually the lagoon may become a salt marsh, then finally be filled in by deposition.

Rather than these first two landscape types, you may instead choose to study the two alternative landscape types mentioned in the examination arrangements. These are upland limestone, and rivers and their valleys.

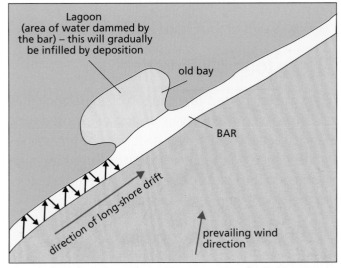

Figure 1.10 A bar is formed as a spit grows across a bay, joining up two headlands

Landscape type 3: Upland limestone

Key points !

 * You should be able to recognise and describe features of an area of upland limestone.
 * For examination purposes these features include: limestone pavements, potholes/swallow holes, caverns, stalactites, stalagmites and intermittent drainage.
 * You should be able to describe and explain the processes that led to the formation of these features.

Introduction

Figure 1.11 shows a summary diagram of the features associated with areas of upland limestone.

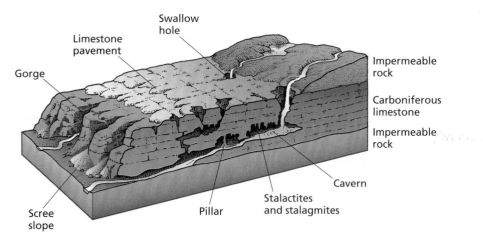

Figure 1.11 Features of an upland limestone landscape

Limestone is a sedimentary rock consisting mainly (at least 80 per cent) of calcium carbonate. Depending on its age it can form several different land forms. Carboniferous limestone was formed about 250 million years ago and the landscapes have specific features that are immediately recognisable.

How were individual features formed?

1 Limestone pavements

When glaciers passed over the top of an upland limestone area, the top-soil was removed leaving an area of exposed rock. The subsequent chemical action of rainwater dissolved the limestone. Joints were widened and deepened on the surface, creating large blocks resembling pavements. The cracks or fissures between the blocks are called **grykes** and the blocks themselves are called **clints**.

2 Potholes/swallow holes

Potholes/swallow holes formed where persistent widening of a major joint occurred, possibly due to a stream disappearing underground.

3 Underground caverns

As the process of dissolving the limestone continued underground, sections of the rock may have collapsed onto the bedrock, creating underground caves. Where the surface water met the impermeable underground rock, this may have led to the creation of underground lakes and streams.

4 Stalagmites and stalactites

Stalactites and **stalagmites** form underground in caverns. Through chemical weathering, particles of limestone are dissolved in solution by rainwater. This water percolates through the rock and drops are deposited on the ceiling and floor of caverns. Gradually the moisture evaporates and the deposits of limestone are left. The deposits of limestone left hanging down from the ceiling of the cave are called stalactites. The deposits that build up from the ground are called stalagmites.

5 Intermittent drainage

Intermittent drainage occurs when streams that drain areas of impermeable rock carry on into a limestone area and disappear through the permeable limestone. This interrupts the course of the stream. The stream flows along the bedding planes until it reaches the underlying permeable rock. Eventually, as the stream flows along the water table, it emerges at the surface at a lower level. This water is called a spring. Areas of upland limestone in Britain form a rolling 'plateau-like' landscape that has no surface drainage. Due to the lack of water, vegetation is sparse or non-existent. Exposed hard, grey limestone is clearly seen on the surface.

Key points !

Stalagmites
* rise from the ground
* have round or flattened ends

Stalactites
* hang from the ceiling
* have pointed ends

A good way to remember this is to focus on the **c** and the **g** in each word:
Stala**g**mite – **g** for **g**round
Stala**c**tite – **c** for **c**eiling

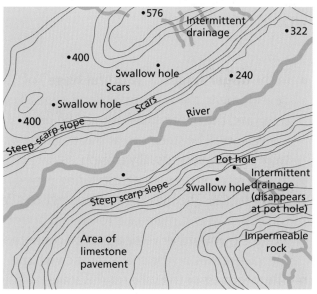

Figure 1.12 Features of upland limestone on an Ordnance Survey map

Example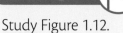

Study Figure 1.12.

    Explain the formation of a limestone pavement.

    You may wish to use diagrams in your answer. **4 marks**

Sample answer

In limestone there are cracks running all the way through it. Since limestone is permeable when rain falls it is absorbed into the rock (✓). But when the rain falls through the atmosphere it picks up fumes from car exhausts and turns it into a weak acid (✓). So when this rain falls on the limestone it dissolves some of it (✓). When this is repeated many times those little cracks grow wider and deeper. The cracks are called grykes and the flat pieces of limestone are called clints (✓). This is how a limestone pavement is formed.

Comments and marks

This answer gained marks for reference to weak acid in the atmosphere, the idea that acid dissolves the limestone resulting in the cracks being opened up and forming clints. The answer gains **4 marks**.

Landscape type 4: Rivers and their valleys

Stages of development and associated features

You should be able to recognise, describe and explain the formation of the main features of a river and its valley in all of the three stages of its development.

River valleys often take different shapes depending on their stage of development.

Upper stage

V-shaped valleys

- In the **upper stage** (or '**youthful stage**') of development, the gradient or slope is usually very steep and the river is fast-flowing. The sides of the valley will be steep and at this stage the main work of the river is usually erosion.
- The erosion process is greatest during periods of heavy rainfall. The river has more energy to affect the bed load, and material is rolled and bounced along. Downward or vertical erosion occurs and the valley takes on the characteristic **v-shape**.
- The steepness of the valley at this stage is also greatly influenced by factors such as local climate, rock type and local vegetation, which can affect chemical erosion. As the river is forced to wind its path due to variations in the relief and rock type, features called interlocking spurs are formed.

Waterfalls

- If a river flows over hard rock and then over a band of less resistant rock, the less resistant rock will be worn away much more quickly than the overlying rock. Eventually this can lead to the formation of a **waterfall**.
- The velocity increases as the river drops, allowing the river to erode the weaker rock further.
- Eventually so much of the underlying rock may be eroded that there is nothing left to support the rock above. The overlying rock then collapses.
- As the process is repeated the waterfall will retreat and this may eventually lead to the formation of a gorge. A gorge is a deep valley with very steep sides and a narrow valley floor.

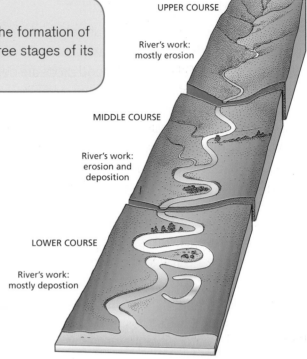

Figure 1.13 Features of a river valley

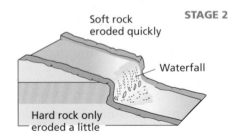

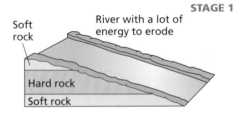

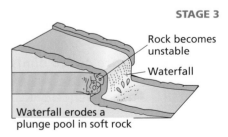

Figure 1.14 Stages in the formation of a waterfall

Middle and later stages

In the middle and later stages the valley sides are less steep and the gradient is gentler. The width of the river increases and bends or **meanders** begin to form.

Meanders

- In the middle stage of the valley, the valley sides are less steep, although they may still be hilly.
- The gradient becomes gentler and the width of the valley increases, with an increase in flat land along the sides of the river.
- River bends or meanders begin to appear as the river finds the course of least resistance. The speed of flow of the river varies across the meander.
- The rate of flow is much slower on the inside bend of the meander. At this point the water deposits material. On the outside bend the velocity is faster and the river erodes the bank. Both of these processes tend to increase the curve in the river.
- In the lower or final stage of the valley, the river widens and flows more slowly across the land. The river floodplain (see below) usually increases in width and large meanders are common.

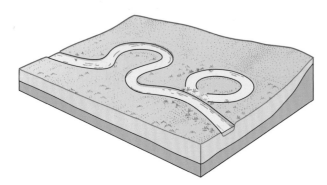

Figure 1.15 A meander and an oxbow lake

Oxbow lakes

- As the size of a meander increases, eventually the river may cut a new channel between the narrowest points of the bend.
- The feature that is left cut off from the river is called an **oxbow lake**.

Levees

- Levees are high raised banks on the side of a river in its lower course.
- They are formed by the river depositing material during times of flood when it overflows its banks.

Recognising and describing river and valley features on an OS map

Some of the questions set on rivers will be based on Ordnance Survey maps. The question may ask you to locate a section of a river by giving you two six-figure grid references. The question may then ask you to describe the main physical features of this section of the river and its valley. If referring to certain physical features, give appropriate four- or six-figure grid references.

- When describing the features of a river and its valley on an Ordnance Survey map, you should refer to:
  - the stage of development of the valley, for example young, mature or old, as determined from the contour patterns of the valley
  - the width of the valley
  - direction of flow, as indicated by both contour lines and spot heights at various points along the course
  - the gradient, i.e. steep or gentle, using spot heights and contours as a basis for deciding on the steepness
  - the width and speed of the river (usually worked out from the gradient)
  - distinct features within the valley, for example steep slopes, meanders, tributaries, oxbow lakes, **braiding**, floodplain width and so on.
- A common error when candidates answer this type of question is to refer to human features such as bridges, roads, houses or industries. These are irrelevant since the question has specifically asked for physical features.

Key words and associated terms

Please see pages 31–32 for key words and definitions on this topic.

Land use related to the four specified physical landscapes

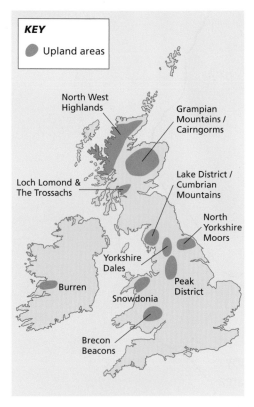

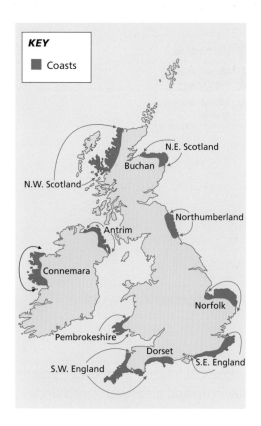

Figure 1.16 Physical landscapes in the British Isles

Key point !

In each of your selected areas (**either** 'Glaciated uplands' and 'Coastal landscapes' **or** 'Upland limestone' and 'Rivers and their valleys'), you should be able to show knowledge and understanding of land uses appropriate to these areas.

General land uses that you should be familiar with in relation to different landscape types include the following.

Farming

There are several different types of farming in Britain and each type has a close relationship with the physical environment. Figure 1.17 shows the distribution of the main farming types throughout Britain.

Climate often affects or limits the type of farming present.
- Crops will flourish if there is the correct amount of rainfall, or die if there is too much or not enough. Different types of crop have different requirements.
- Temperature changes throughout the year determine growing seasons.

- Farmers may opt for livestock farming as opposed to crop farming (or vice versa) due to the influence of climate.
- The flatness of land, steepness of slopes and relative fertility of the underlying soil also encourage certain types of farming and limit others.

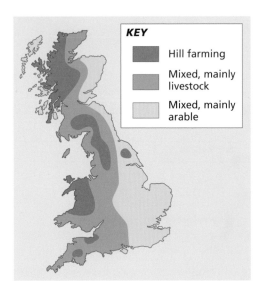

Figure 1.17 Distribution of main farming types throughout Britain

Forestry

Large plantations of commercial forestry are found throughout many parts of Britain, especially in upland areas with steep slopes and in areas where soil quality may be relatively poor. This is not so much because forestry is attracted to such areas but because in areas where physical conditions are better, the land will be too valuable to use for forestry. This land may be more suited to settlement, industry or farming. Over 75,000 tons of timber is produced from commercial forestry each year.

Since the 1920s, the **Forestry Commission** (or Forest Enterprise as it has been known) has planted large areas of Britain with coniferous trees such as pine, fir, spruce and larch. A total of 3.1 million hectares of land in Britain has been planted with trees grown for commercial forestry.

Industry

Industrial land use can vary from primary industries such as mining and quarrying, to large, heavy industrial complexes such as iron- and steelworks, to light industrial estates. It may include power stations such as coal- and oil-fired plants, and hydroelectric power schemes in upland areas.

Recreation and tourism

- Some recreational pursuits, such as golfing, sailing, hillwalking, climbing, sightseeing and observing nature, are closely linked to the nature of the physical environment.

- Tourism is one of the most important service industries in Britain. The industry caters for the needs of hundreds of thousands of people who wish to visit places for the purpose of recreation and leisure.
- In Britain there are many different types of tourism enterprise located in both countryside and urban areas, ranging from city breaks to National Parks and other areas of outstanding natural beauty. Figure 1.18 shows the location of National Parks – areas of outstanding scenic beauty throughout Britain. These have become magnets for tourists.

Figure 1.18 UK National Parks and main centres of population

Water storage and supply

Many of the lochs and lakes in Britain are used for water storage and the supply of water to towns and cities.

Renewable energy

Hydroelectricity

Hydroelectric schemes are one important example of renewable energy projects. They generate electricity that is fed into the National Grid. These schemes may involve flooding valleys in highland areas. Alternatively, they may use natural features such as corries, which may already be filled with lakes or tarns. Water flows down the mountain through pipes and is used to turn turbines at the foot of the mountain. These turbines generate the electricity. These schemes are often built in remote areas, where few people can see them, so as to protect the natural scenic beauty of the landscape.

Wind farms

● Wind farms are fast becoming a familiar feature of rural and coastal areas. They are designed to harness the wind in order to generate energy.

Key words and associated terms

Please see pages 31–32 for key words and definitions on this topic.

Conflicting land uses and strategies to manage them

Key point !

With reference to **one** of the two landscape types you have studied, you should be able to show knowledge of the conflicts that can arise between various land uses within this landscape.

Competition for land is perhaps the major cause of conflict between land users in all of the landscapes studied.

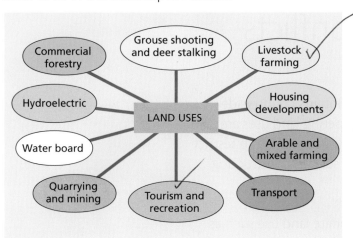

Figure 1.19 Competition for land use

- In the examination at National 5, you may be asked to give reasons why certain land uses are in conflict with each other. Usually you will be asked to select two or more land uses from a given list or diagram.
- Farming requires substantial areas of land and all other land uses, such as settlement, industry, forestry and reservoirs, obviously limit the amount of land available.
- Industrial and **settlement** land uses take up large areas of the countryside. Cities are always growing outwards and often begin to encompass rural areas, including small rural settlements.
- Apart from competition for land, some land uses can be in direct conflict with each other because their activities directly affect or threaten one another. Agriculture and tourism are a good example of this.
- Land use that produces any kind of pollution, such as fumes, litter, chemicals or noise, will have a detrimental effect on crops and livestock. Industry, settlement, communications and tourism will often be in direct conflict with agriculture for this reason.

Tourists in any rural area can seriously affect other land uses, either directly or indirectly. For example, large numbers of tourists require certain facilities such as roads and other services. Traffic issues such as road congestion and traffic fumes can affect natural vegetation and disturb the peace and quiet of the rural environment.

Large numbers of tourists can threaten farming in a variety of other ways, including soil erosion, leaving litter, pollution of rivers and lakes, dogs worrying sheep, leaving gates open and trespassing on farmland. Similarly, forested areas may be endangered by tourists/campers being careless with fire or damaging footpaths and leaving litter.

As well as being able to identify when and where conflicts exist, you must also be able to describe the kinds of measures taken to resolve them, either at local or national level.

There are certain government agencies, at local and national level, whose task it is to monitor and, wherever possible, prevent or resolve major conflicts. These include National Park and **Country Park** Authorities.

Management strategies and solutions to deal with land use conflicts

General points

- National Park Authorities help to ensure that Britain's countryside is protected and that conflicts caused by competing land uses are resolved. Occasionally this can involve employing legislation to restrict access to certain areas, or protective measures such as constructing footpaths to guard against erosion.
- Other measures used include:
  - planning legislation that prohibits inappropriate land use such as housing/industry
  - zoning of tourist areas
  - providing education centres for visitors and visitor centres.

National Park Authorities also employ rangers to patrol the parks. The task of these rangers is to monitor potential problems caused, for example, by tourists and take action to resolve these problems. These strategies have often been very successful in ensuring that tourism is sustainable and that the natural environment is protected and conserved.

In addition to National Park Authorities, there are a number of public and voluntary bodies that have a major role in protecting and conserving the countryside in Britain. These include the National Trust, the **Countryside Commission,** Country Park Authorities, Greenpeace, the National Wildlife Trust, Coastal Protection Agencies and a variety of local pressure groups that take an active role in protecting the natural environment.

Their efforts to conserve, protect and manage the problems of scenic areas under threat include:
- raising public awareness of environmental issues through sponsored education programmes

- purchasing land in order to control the land use and protect the area from environmental misuse
- raising the profile of local issues over land use through articles in local newspapers and information pamphlets and leaflets
- protesting in various ways, such as through public meetings and demonstrations, against measures that could harm the environment.

Impact of strategies

These organisations often target issues such as industrial and traffic pollution and congestion in rural areas, the construction of new motorways through Country Parks, and so on. Again, these bodies have often been very successful in their efforts to maintain and conserve the natural environment in various areas. Their work has often led to public enquiries into proposed developments that could have a detrimental effect on the environment, such as the building of nuclear and hydroelectric power stations, industrial estates, new motorways and similar projects.

Key points ⓘ

* In class you may have studied several examples of different landscapes in the UK in order to demonstrate your knowledge of this topic.
* For the purposes of this text, three sample areas have been selected for brief case studies. These include:
  - the Lake District (glaciated uplands)
  - the New Forest coastline (coastal landscape)
  - the Peak District (upland limestone).
* These sample areas may not be the same as those selected for your own study areas. However, they will provide an example to help you answer questions on areas that you have studied.

Case Study 1: The Lake District (glaciated upland)

Background

Three main rock types make up this area, namely igneous (e.g. granite), metamorphic (e.g. slate) and sedimentary (e.g. grits and limestone). The Lake District contains England's best example of an upland glaciated area, with a wide variety of features including pyramidal peaks, tarns, ribbon-shaped lakes and u-shaped valleys. The most famous lakes in the region include Windermere, Ellesmere, Ullswater, Coniston and Thirlmere.

Land uses

All of the following land uses are located in this area.

Farming

The area is unsuited to crop farming. The steepness of the slopes make it almost impossible to use machinery such as combine harvesters, and the cold temperatures and high rainfall limit the growing season and affect soil fertility. The only type of farming that is feasible is hill sheep farming, with cattle occasionally being raised on lower, less steep land.

Forestry

Large plantations of coniferous forests are found in the Lake District. This activity is well suited to this area of steep slopes, poor soils and relatively inhospitable climate. The trees also protect the slopes from soil erosion.

Industry

Due to the lack of flat land, little manufacturing has been attracted to the area. The main type of industry is quarrying for granite and slate for roads and roofs. Limestone is also quarried, for use in steelmaking elsewhere. However, the number of quarries operating has been significantly reduced in recent years.

Water supply

The Lake District has supplied Manchester with water for over 100 years, despite being over 150 kilometres away from the city. The lakes are natural reservoirs in an area of high rainfall and are much more economical for Manchester to use than building artificial reservoirs would be. The lakes supply up to 30 per cent of the water needs of this part of Britain.

Recreation and leisure/tourism

The Lake District is very attractive to tourists. It offers a variety of physical attractions such as the mountains and lakes, and is an ideal location for activities such as hillwalking, mountain climbing, adventure holidays, water sports, fishing and general sightseeing. Over 12 million people visit the Lake District each year and this number is increasing annually. Tourism is very important to the region economically. Recent developments in the region have included extensions to hotels and leisure complexes, timeshare complexes, marinas and cable cars/ski lifts.

Fifty per cent of visitors are either from regions immediately next to the Lake District, such as Newcastle, Manchester and Leeds, or from areas linked to it by the M6, such as Birmingham. Ninety per cent of visitors travel there by car. Access to the Lake District has been made easier through the construction of motorways such as the M6, bringing many more visitors from the south of Britain to the area.

The area was designated a National Park with the main purpose of offering city dwellers a place to escape from the city and enjoy the benefits of a protected countryside. The environment has also benefited from National Park status. Efforts are made to protect the physical environment by the **National Park Authority** and other bodies.

However, housing is in short supply in the area and house prices have increased dramatically. Much of the existing housing is now used for 'second homes', which means that local people, especially young people, are forced out. This causes resentment among the local population. Local businesses also suffer due to lost trade when second homes are left empty for much of the year. Tourism has had a significant impact on the natural environment with increases in pollution, traffic congestion, footpath erosion, and changes in the traditional rural character of many villages.

Land use conflicts in the area

Due to competition for land, there is conflict between most of the major land uses in the Lake District. One major area of conflict exists between tourism and farming. This is due to factors such as:

- increased traffic congestion, due to tractors holding up traffic and heavy use of small rural roads by tourist traffic
- damage caused by tourists to farms, through leaving litter, trespassing in fields, increased pollution, gates left open, animals worried by family pets, and so on
- farmers have sometimes blocked access to public footpaths.

There is also conflict between farming and other major land users, especially water boards who flood areas of valuable farmland and developers looking for land for industrial/housing purposes.

The increased use of the lakes for water skiing, power boating and jet skiing creates conflict between people enjoying these activities and other lake users, such as swimmers, sailors, anglers and those interested in wildlife.

Management strategies to deal with land use conflicts

- Legislation such as the Greenbelt Act is enforced to protect the area and conserve it against industrial and urban developments.
- Some strategies involve partnerships between the National Park Authority, the National Trust and the tourist and hotel industries to encourage sustainable tourism. This involves:
  - raising visitors' awareness of responsible tourism
  - raising money to restore and conserve the landscape
  - ensuring that tourism and conservation work together to benefit the local community.
- Strict planning laws are observed to ensure that any development is both in character with the area and environmentally sustainable.

Case Study 2: The New Forest (coastal landscape)

Background

The New Forest is an area of southern England that incorporates the New Forest National Park. The region is known for its heathland, forest trails and native ponies. The park's abundant plant and wildlife includes owls, otters and wolves, along with a diverse mix of fungi and woodland flowers. The New Forest was proposed as a UNESCO World Heritage Site in June 1999, and it became a National Park in 2005. The National Park receives an estimated 17 million day-visits a year. It is thought that these visits generate approximately £120 million in annual income and support more than 2,500 jobs. The park's most popular recreation activities include walking, cycling, horse-riding, visiting its tearooms and cafés, and sightseeing.

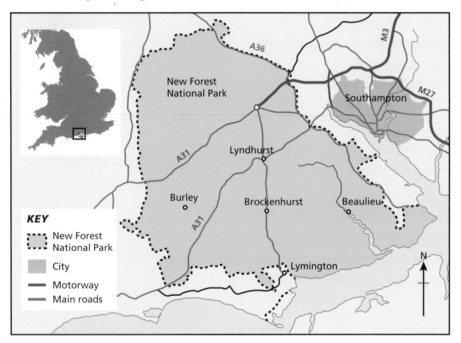

Figure 1.20 The New Forest coastal area

Land uses

There are a wide variety of land uses present, including: farming, industry, forestry, tourism, and recreation and leisure.

- The **estuaries** of the Lymington and Beaulieu Rivers are centres for sailing, boat building and boat repair.
- Much of the area of Southampton Water directly east of the New Forest coastline is developed for major industry such as the Esso oil refinery and petrochemicals complex, housing developments and power stations.
- Coastal marshes in the area are used for nature conservation, despite the risk of water pollution caused by industrial waste and domestic sewage.

17 million tourists are attracted to the area annually, and there are over 200,000 educational visitors (including students and school pupils)

catered for every year. Economically the area benefits from the wide range of economic activities including tourism, industry, forestry and farming. These different activities have a major impact socially, economically and environmentally on this area of coastline.

Land use conflicts in the area

The area is a magnet for population, since it is very attractive to tourists and also as a place to live, work and retire. Activities such as tourism and industry generate a great deal of money for the people who live and work in the area.

However, the environmental effects of tourism, port and ferry services and increased traffic congestion include footpath erosion, increased demand for land for car parks and other amenities, threats to wildlife habitats, and greater risk of marine pollution. There are conflicts between fishing, water sports, marine archaeology and the presence of the UK's longest onshore oilfield. All of these activities combine to threaten the natural environment and to change the natural balance and ecological diversity of the area. In addition, the coastline is under threat from natural forces such as waves, currents and groundwater movement.

Management strategies to deal with land use conflicts

A document called *New Forest 2000*, outlining the strategy for sustaining and improving the quality of the environment in the region, was published in 1990. Subsequently, the Recreation Management Strategy (RMS) was set up to plan the management of outdoor recreation in the New Forest National Park from 2010–2030. This Strategy sets out a framework for the management of outdoor recreation in the New Forest National Park over the coming 20 years. Its main purpose is to balance the needs of the various land users and manage land-use conflicts. The original National Park Management plan was updated in 2015.

- Measures outlined in this strategy included efforts to:
  - reduce pollution levels
  - protect the area's scenic beauty
  - improve the appearance of the coast
  - maintain the economy
  - protect the coastline
  - educate the public
  - conserve features of historic and archaeological interest.
- The coastline needs to be managed in order to:
  - sustain human activities in the face of the threat of marine erosion
  - preserve coasts for conservation reasons
  - protect coasts from uncontrolled development.

- Responsibility for managing coasts generally lies with three agencies: the Environment Agency, DEFRA (Department for Environment, Food and Rural Affairs) and district councils.
- Coastal defence strategies include: seawalls, the use of large, irregular rocks, which protect the coastline from wave erosion, gabions (wire baskets filled with rubble), groynes and embankments.

Impact of strategies

Each of these measures has its own advantages and disadvantages, not least of which is cost. Some are relatively cheap whereas others, such as seawalls, can be very expensive to build and maintain. There have also been attempts to protect coastlines from flooding using dykes and flood walls. Nature reserves have been created to protect wildlife and two Country Parks have been created to encourage sustainable tourism. These strategies are managed by public authorities, which operate the nature reserves and the Country Parks. These efforts in the New Forest area have met with considerable success.

Case Study 3: The Yorkshire Dales (upland limestone)

Background

The Yorkshire Dales are located on the eastern side of the Pennine Hills. They consist mainly of two rocks types, carboniferous limestone and millstone grit. The region contains the largest area of upland limestone in the UK. All of the landscape features discussed in the section on upland limestone can be found in this area.

Land uses

Due to difficulties presented by the physical landscape and local climate, there are restrictions on possible land uses. The main land uses in the region are hill sheep farming, quarrying, water storage, forestry, military training areas, and areas used to produce renewable energy such as wind farms.

Limestone is in great demand in the building industry. With an annual income of over £6 million, the eight quarries in the area provide work for over a thousand employees.

Many tourists are attracted to the area due to the scenery. With over 2000 km of public paths, it is not surprising that hillwalking is a very popular activity in the area. The area is also popular for its underground features and attracts potholers and cavers. The tourist industry provides a boost to the local economy by attracting an average income in excess of £30 million and employment for thousands of people.

Forestry exists only in the millstone grit areas, since trees cannot grow well in the dry limestone areas.

Due to the permeability of the local rock, water storage schemes are severely restricted.

The area is not heavily populated and this makes it suitable for military training purposes, since there are few people to be disturbed by this land use.

Land use conflicts

There is conflict between all the major land users, such as hill sheep farmers, quarries, tourists, military training and housing. For example, quarries are unsightly and spoil the natural scenery for tourists. The lorries used to transport the limestone are noisy and quarrying also creates pollution. Hillwalking has potential for eroding the landscape. Traffic congestion caused by increased tourism is a source of conflict between the local population and the visitors. These activities do impact on the environment to some extent but the impact is not felt as much as in other parts of Britain, such as the Lake District, where economic development poses much more of a threat.

Management strategies to deal with land use conflicts

The area was designated a National Park in 1954.

The National Park Authority employs many of the same measures as in the Lake District to protect and conserve the natural environment. There are planning restrictions on the building of new housing or industrial developments that may adversely affect the human and natural environments. Farmland is protected from the impact of tourism and other activities by the provision of footpaths, restricted access to certain areas, and greenbelt legislation to limit new housing developments and industrial land use.

Key words and associated terms

Abrasion: The process by which rocks within ice sheets and rivers scrape and erode the land over which they pass.

Alluvium: The material deposited by a river, usually over its floodplain.

Arête: A narrow ridge between two corries, created as the corries are formed on two adjacent sides of a mountain.

Braiding: The process by which rivers divide into separate channels through material being deposited by the river mid-stream.

Clint: block of limestone forming a limestone pavement

Corrie: An armchair-shaped hollow on the side of a mountain. They form by ice filling a hollow and eroding the side of the mountain through abrasion and plucking and by rotational movement at the base of the hollow. When the glacier melts a lake or loch may be left, called a corrie-loch or tarn.

Country Park: An area in the countryside surrounding a town or city that has been set aside for people to visit.

Countryside Commission: This organisation was set up by the government to monitor countryside areas and protect them from harmful development.

⇨

Erosion: The process by which rocks and landscapes are worn away by agents such as moving ice, wind, flowing water and sea/wave action.

Estuary: The mouth of a river where it meets the sea.

Forestry Commission: This organisation is responsible for planting and looking after forests throughout the UK.

Glacier: A large mass of moving ice that changes the shape of the land over which it is passing.

Grykes: Deep joints or fissures on the top of limestone pavements. They are formed through the chemical reaction of rainwater and limestone. When deep enough they split the surface to form large blocks called **clints**, giving the 'pavement'-like appearance.

Land use conflict: This occurs when different activities compete with each other to make use of the land, for example farming and tourism.

Meanders: Bends formed in the middle or lower courses of rivers.

National Park Authority: The organisation that looks after Britain's National Parks, which are areas set aside throughout Britain for the recreation and enjoyment of the public. It also aims to protect these areas of outstanding scenic beauty.

Oxbow lake: Lake formed from former meanders when the loop of the meander is cut off from the main channel as the river finds a new, more direct path.

Plucking: The process by which moving ice tears rocks from the surface over which it moves.

Pyramidal peak: A jagged peak on top of a mountain.

Settlement: A place where people live.

Spit: A long, narrow piece of land made of sand or shingle, jutting out into the sea, formed by marine deposition.

Stalactites and stalagmites: Structures formed underground in limestone areas. Calcium carbonate in solution drips from the ceiling of underground caves and is deposited on the floor. The moisture evaporates leaving behind small deposits of calcium carbonate, which gradually build up to form stalactites (from the ceiling) and stalagmites (on the floor). These can eventually meet to form limestone pillars.

Transportation: The process by which rock particles are carried by rivers, glaciers or wind.

Truncated spur: A piece of land, the bottom of which at one time jutted into a valley and was cut away or eroded by a glacier flowing through the valley.

U-shaped valley: A valley with very high, steep sides and a wide, flat bottom formed by a glacier flowing through the original valley. It is usually occupied by a small river, which is known as a 'misfit stream' since it was not the original river flowing through the valley.

V-shaped valley: A narrow valley typical of the **upper** or **youthful stage** of river development.

Waterfall: A steep break in the course of a valley, often associated with changes in rock type along the course of a river.

Weathering: The process by which rocks are worn away. This may be through physical action, such as flowing water or the wind, or through a chemical reaction between rocks and rainfall, which may have become acidic.

Development indicators

Economic indicators

Economic indicators may include:
- **Gross National Product (GNP)**: the total amount of money from goods and services produced by a country in one year
- percentage of working population (known as the **active population**) employed in agriculture or industry
- average per capita income
- consumption of electricity per capita (kilowatts per capita)
- percentage unemployment
- data showing steel production in tonnes per capita
- trade patterns in terms of import and export figures
- trade balances in terms of surplus or deficits.

Social indicators

Social indicators may include:
- **birth rates**, **death rates**, **infant mortality** rates and **life expectancy** rates
- **population structure** in terms of the distribution of age and sex within the population
- average calorie intake per capita
- the average number of people per doctor
- literacy rates as an indication of the level of education
- the percentage of the population with access to clean water
- the percentage of homeless people
- the percentage of the population attending primary/secondary school.

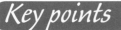

Decisions on levels of development should be based on a number of indicators used together, rather than on individual indicators such as Gross National Product (GNP) alone. This is because:

- The use of single indicators can be misleading, since the data is based on averages and does not reveal the whole situation.
- For example, average per capita income or GNP per capita for Saudi Arabia may seem high and suggest a high level of development. However, income distribution is very unequal, varying from extremely high for some to very low for others.
- Similarly, GNP may indicate a high level of development but be based on a single commodity such as oil.

A much more accurate picture of the stage of development of a country is given by combined indices based on several different indicators. These include the Physical Quality of Life Index (PQLI) and the Human Development Index (HDI).

The social and economic indicators discussed earlier can be used in combination to produce maps showing varying levels of development throughout the world (see Figure 2.1). Tables and diagrams based on this data can also depict countries at different stages of development.

In general, areas with low Gross National Product, high birth rates, low life expectancy rates, low literacy rates, low income per capita and a high percentage of the workforce working in agriculture would be best described as developing countries. The opposite situation applies in developed countries.

Accuracy of development indicators

As noted earlier, the use of one individual development indicator can be misleading, especially if it is based on average figures, for example income per capita or GNP. This does not reveal the possible wide variations within the same country, with some people being very wealthy while others live at subsistence level. In some Middle Eastern oil-producing countries, for example, GNP might appear to rank alongside those of highly developed countries but this wealth is not evenly spread throughout the population. Therefore, using a combination of indicators to produce a 'quality of life index' is the best method of assessing levels of development in any given country.

KEY
- PQLI of 90 or above
- PQLI of 78–89
- PQLI of 56–77
- PQLI of 31–55
- PQLI of 30 or below

Figure 2.1 Levels of development based on the Physical Quality of Life Index

Population structure

- The structure of the population of a country is defined in terms of age and sex distribution. Males and females are subdivided into different age groups, for example 0–4, 5–9 and so on up to the 80+ category.
- The number of males and females in each age group as a percentage of the total population is plotted on a graph called a population pyramid.
- Structures can indicate different levels of development.
- Analysis of population structure graphs (age/sex pyramids) can reveal patterns of birth and death rates and an estimation of life expectancy.

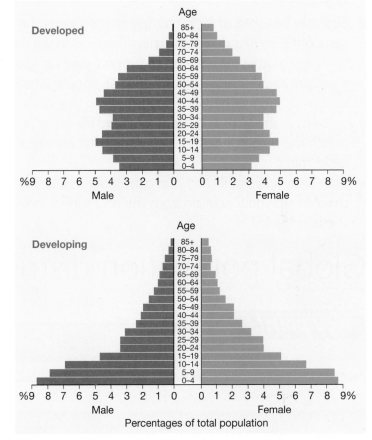

Figure 2.2 Typical population structures for a developed and a developing country

Population structure in a developed country

As Figure 2.2 shows, in developed countries:
- the birth and death rates are both fairly low
- people have a fairly high life expectancy
- the number of people of both sexes in the groups beyond 60+ is a relatively large proportion of the rest of the graph.

The structure is fairly well balanced since the number of males and females is similar.

Population structure in a developing country

- This pyramid has a wider base, which indicates a high birth rate.
- The shape narrows for both males and females until it is very narrow in the 65+ age groups.
- This indicates that life expectancy is low.

Skills involved in the interpretation of population data

- Questions on population data may ask you to examine graphs, diagrams, tables or maps and draw conclusions. For example, you may be given a table showing various measurements of population for two or more countries. It might contain details of life expectancy, birth and death rates, infant mortality rates and other indicators such as medical provision.
- You may be asked to explain differences between the countries. You can do this by first describing the differences in the data, and then referring to other information to support your conclusions. Similarly, if you are asked to use diagrams such as population pyramids to explain differences between countries, you should first identify any patterns in the birth and death rates and life expectancy. You can then match them appropriately to countries that are either developed or developing.
- Your knowledge of factors that contribute to levels of development should help you to explain both the structures and the reasons for them.

Global population distribution

> ## Key point ❗
>
> What physical and human factors influence global population distribution?

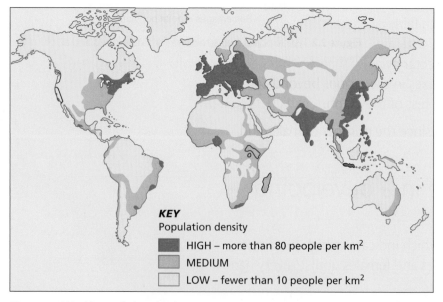

KEY
Population density

- HIGH – more than 80 people per km^2
- MEDIUM
- LOW – fewer than 10 people per km^2

Figure 2.3 World population density

Physical factors that influence population distribution and **population density** include the following.

Relief

Relief refers to the height and shape of the land.

Positive factors

- Flat land for building settlements.
- A good supply of water from rivers and valleys.
- Easy access, for example location in a valley or on a coastal plain.

Negative factors

- Situations that are fairly inaccessible, such as mountains, jungles or deserts.
- Situations that are remote from other good settlement land.
- Situations that are difficult to build on, for example steep slopes.

Climate

Climatic factors that influence population distribution and density include the following.

Positive factors

- Climates that are suitable to live and work in – temperate climates that have mild to warm temperatures throughout the year and moderate/ adequate amounts of rainfall for water supplies are attractive.
- Climates that are suitable for growing crops.
- Climates that are suitable for the development of tourism, such as Mediterranean climates, which are warm and dry in summer, mild and wet in winter.

Negative factors

- Areas where the climate is inhospitable, for example too hot, too cold, too wet or too dry.
- These include climates such as hot deserts, tundra, Arctic and some equatorial/tropical climates.
- These areas generally have low population densities.

Resources

Natural resources are a source (or a potential source) of wealth. If these are accessible in an area, large numbers of people are usually attracted to them.

Natural resources include:

- good fertile soil for farming.
- flat land for arable farming.
- areas which have large populations and where labour is the main input into farming. Such areas include places such as south-east Asia where peasant farming is the most common type of agriculture.
- mineral resources such as oil, coal, ores, and precious and semi-precious minerals such as gold, silver, copper and bauxite. These resources attract both primary and manufacturing industry and consequently large numbers of people to work in these industries.

Areas that lack natural resources do not offer much opportunity for human development and settlement. Such areas are typified by low population density.

Employment opportunities

- Due to a combination of good climate conditions, suitable relief and the presence of natural resources, many areas of the world were highly suitable for industrial and agricultural development. Such development requires large numbers of people for labour. In developed countries most of the workers are employed in industries in towns and cities.
- In developing countries the majority of workers are usually employed in some form of farming. As agriculture and industry developed, more jobs were created attracting more people from other areas.
- In developed countries the majority of industrial jobs are in the service sector. In developing countries most industrial jobs are in primary industries such as mining, forestry, fishing and quarrying.

If asked to explain patterns of population distribution:
- You should describe where the main areas of population occur, or perhaps where the density of population is high, medium or low.
- You can explain world population distribution by referring to the comparative influence of physical and human factors.
- Factors that attract and repel population are called 'positive' and 'negative' factors, respectively.

Example

Study Figure 2.3.

Explain why some parts of the world are more densely populated than others.

Refer to both physical and human factors in your answer. **6 marks**

Sample answer

Some parts of the world are more densely populated than others because some areas have plenty of resources like gold, bauxite and coal (✓) and these allow industries to develop which attract people to jobs (✓). Good flat land is good for building on and for farming producing food (✓). Good climates attract people. Cold areas like the Tundra have fewer people as it is difficult to produce food in the extremely cold temperatures (✓). Rain forest areas are too hot and wet to maintain a large population as well as insects which cause diseases like malaria (✓).

Comments and marks

The references to resources, giving an example as well as an explanation, gains marks. Explaining why flat land is good gains a mark. 'Good climates attract people' is a description so gains no marks. The next two points explain the climate so gain marks. A better example would explain climate in more detail. This answer gains **5 marks out of 6**.

Population changes

Birth rates (crude)

This figure indicates the number of people born in any given year, per thousand of the population. Since this is the basic measure it is termed 'crude'.

Death rates (crude)

This indicates the number of people who die in any given year, per thousand of the population.

Natural growth rate

Subtracting the death rate from the birth rate gives a basic indication of the amount by which the population is increasing (or decreasing) each year, per thousand of the population.

Average life expectancy

This figure indicates the average number of years a person can expect to live within any given country, for example males 67 years, females 70 years.

Infant mortality rate

This indicates the number of deaths of children under the age of one year per thousand of the population per year in a country.

Factors affecting birth rate ☑

Factors leading to high birth rates include:
- religious beliefs that prevent use of artificial methods of birth control
- lack of access to birth control
- lack of education
- use of children as labour to increase family income
- large families to support parents in old age
- traditional/cultural reasons for large families
- possible insurance against high infant mortality rates.

Factors leading to low birth rates include:
- people marrying and having children later for economic reasons
- widespread availability and use of birth control methods
- women putting their career before having children
- education on birth control being widely available.

Factors affecting death rate

Factors leading to high death rates include:

- widespread poverty
- poor diets/malnutrition
- widespread disease due to poor healthcare/hygiene/sanitation
- lack of access to clean water supply
- lack of medical drugs/hospitals/clinics/doctors
- unhealthy environments (poor housing etc.)
- natural disasters – droughts/floods/earthquakes

Factors leading to low death rates include:

- economic prosperity
- availability of good healthcare, hospitals, doctors and medicines
- plentiful supply of food and well balanced diets
- good housing/sanitation/clean water available
- safe natural environments
- high standards of education/health education.

Other factors

Other factors affecting population growth rates in a country include:

- high/low rates of immigration/emigration
- the level of economic development in a particular country.

Demographic Transition Model

The changes in population growth rates and the effect on population can be shown by the **Demographic Transition Model**. It shows population change over time. There are five main stages to the model. These changes and the reasons for them are shown in Figure 2.4 below.

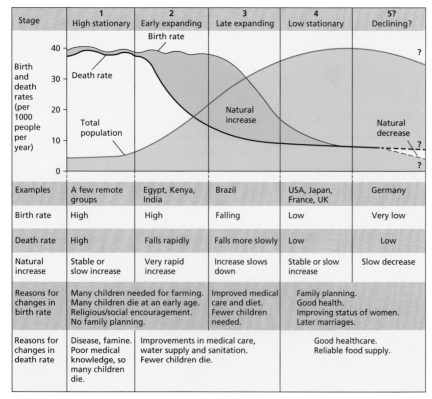

Stage	1 High stationary	2 Early expanding	3 Late expanding	4 Low stationary	5? Declining?
Examples	A few remote groups	Egypt, Kenya, India	Brazil	USA, Japan, France, UK	Germany
Birth rate	High	High	Falling	Low	Very low
Death rate	High	Falls rapidly	Falls more slowly	Low	Low
Natural increase	Stable or slow increase	Very rapid increase	Increase slows down	Stable or slow increase	Slow decrease
Reasons for changes in birth rate	Many children needed for farming. Many children die at an early age. Religious/social encouragement. No family planning.		Improved medical care and diet. Fewer children needed.	Family planning. Good health. Improving status of women. Later marriages.	
Reasons for changes in death rate	Disease, famine. Poor medical knowledge, so many children die.	Improvements in medical care, water supply and sanitation. Fewer children die.		Good healthcare. Reliable food supply.	

Figure 2.4 Demographic Transition Model

Key words and associated terms

Active population: That section of the population of a country which is economically active/working.

Birth rate: The number of births per thousand of the population in a country in a given year.

Death rate: The number of deaths per thousand of the population in a country in a given year.

Demographic Transition Model: Shows the different stages a country goes through towards development.

Developed countries: Sometimes referred to as 'more economically developed countries' (MEDCs). They include countries that have a high standard of living or high physical quality of life.

Developing countries: Sometimes referred to as 'less economically developed countries' (LEDCs). The population generally has a low **standard of living**.

Gross National Product (GNP): The value of all goods and services produced by a country in a given period of time, such as one year. It is used as an indicator of the wealth of a country. However, it does not always reveal how well spread the wealth is among the population in general.

Infant mortality: The number of children below the age of one year who die each year, per thousand of the population.

Life expectancy: The average age a person can expect to live to in a particular country. This is a good indicator of level of development, since people in more developed countries tend to live longer due to better healthcare, better diets, higher standards of education and housing, and so on.

Population density: The average number of people within a given area, for example 100 per square kilometre.

Population structure: The grouping of the population of a country by age and sex. Inspection of the structure may also indicate trends in birth and death rates, life expectancy and the possible impact of factors such as war and migration on the population.

Standard of living: The level of economic well-being of people in a country.

Land use zones

Figure 2.5 shows a summary of the main zones within cities. When these zones are looked at together with a graph showing land value, you should be able to detect a direct link between land use and land value.

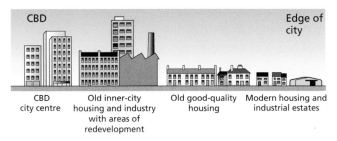

Figure 2.5 Land use zones

The following sections discuss the main features of different land use zones in a typical city and how you can identify them on an Ordnance Survey map.

Zone 1: The Central Business District

- Those functions that provide important services for residents of a settlement are usually located at or near the centre, for example, shops, banks, transport and entertainment. This area is called the **Central Business District (CBD)**.
- You may be asked to identify this area on an OS map by indicating certain map squares. You may also be asked to explain your choice. In your answer you should discuss the types of buildings, the pattern of streets (usually narrow or in a gridiron arrangement), the fact that major roads converge there giving greater accessibility, and the presence of bus and railway stations and public buildings such as town halls.

The **inner city** is made up of the old nineteenth-century industrial and housing zones. Zone 2 and Zone 3 make up the inner city. The inner city is located near the centre of the city and in recent years it has undergone great change and regeneration.

Zone 2: Industrial zone

- Old industrial areas with factories (works) are usually located close to the CBD.
- Accessibility is normally the main reason for this; it allows factories to bring in their raw materials and distribute their finished products easily.
- It also allows workers to get to their workplace easily.
- This zone can be identified on an OS map by the shape of the buildings (usually large blocks) and words such as 'works' printed beside buildings.

- This zone is normally serviced by main roads and railway lines.
- In older cities, this zone may contain heavy industries or may have gap sites in which old industries that have since closed were once located.

Zone 3: Low-cost housing

- Since the area around old factories is not the most pleasant to live in, housing around this area is usually the cheapest. It would have originally been built for workers and their families.
- On an Ordnance Survey map the streets may form a gridiron pattern, with narrow streets.
- The density of housing will usually be high, consisting mostly of tenements.
- Population density will also be high in these areas and many of the houses may be fairly old.
- There are areas with derelict buildings and nineteenth-century housing that have since been demolished and turned into brownfield sites.
- Some of these sites have been replaced by warehouses, new housing, roads and railways.

Zones 4 and 5: Medium- and high-cost housing

- People who could afford to would have begun to move further away from the town centre. As a result most towns and cities have different qualities of residential areas, namely low-, medium- and high-cost.
- In Zone 4, streets may have a curvilinear pattern with cul-de-sacs. Houses may consist of a mixture of tenements, terraced houses and semi-detached houses.
- House prices will be higher than those in Zone 3 and many homes will be owner-occupied.
- Zone 5 may have a mixture of detached and semi-detached villas and bungalows. House prices are more expensive with higher costs. The streets may be wider and the density of housing will be much less than in any other residential zone.
- On the fringes or outskirts of cities and towns there are areas that have been used for newer, more modern industries and out of town shopping areas.
- Some cities, especially in Scotland, also have council house schemes with lower-cost housing built in these areas.

Zone 6: Rural/urban fringe

The rural/urban fringe is found at the edge of the city and marks the boundary between the built-up area and the countryside. This land attracts many different land uses such as housing, trading estates and recreation, which can lead to land-use conflicts.

- On an OS map these settlements will be located outside the main built-up area and may be linked to it by an A-class road or motorway.

The CBD — Walnut st james centre
whip
✓ ## Recent developments in CBDs

The main developments include:

- The building of undercover shopping centres.
- The closures of city centre shops and other businesses due to a fall in the number of customers.
- Some warehouses and former public buildings have been converted into modern flats.
- Many have closed and moved to more affordable premises, often in out-of-town shopping centres.
- There have also been improvements and changes to roads and streets to ease problems of **traffic congestion**.
- The introduction of **pedestrianised zones**.
- Congestion charges, where drivers have to pay a charge to drive in central zones, are now used to deter drivers from taking their cars in the city centre.
- Improved public transport, for example, bus lanes to keep the buses running on time. More buses and trains as well as WiFi to encourage people to travel by public transport.
- New office blocks have replaced older buildings such as bus and railway terminals.
- To encourage population back to the city, buildings such as warehouses and former public buildings have been converted into modern flats.

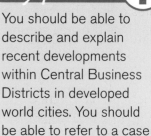

> **Key point** !
>
> You should be able to describe and explain recent developments within Central Business Districts in developed world cities. You should be able to refer to a case study in your answer.

 ## Impacts

- Discouraging people from coming into the city can create problems, especially for shopkeepers and other businesses that rely on such customers for their existence.
- Offering alternatives such as out-of-town shopping centres may help to solve traffic problems, but these centres often lead to the decline of the Central Business District, from which a large proportion of local government finance is obtained.
- Authorities have to balance their solutions very carefully, and national government policies recognise this.

Recent developments in the inner city ✓

The inner city is sometimes called the 'zone of transition'. This is because the old original structure is replaced by a more modern landscape. This is due mainly to the decline and disappearance of the old industries. Their closure left the inner city with many derelict buildings and run-down, polluted areas. This also created a high unemployment rate. To improve the situation, a programme of urban **regeneration** has been ongoing and the area has undergone many changes in recent years.

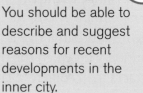

> **Key point** !
>
> You should be able to describe and suggest reasons for recent developments in the inner city.

Urban-decay renewal

In the 1970s, a programme of urban **renewal** began in Glasgow's inner city. The tenement housing in areas like Bridgeton and Dalmarnock was renovated, along with the building of some new housing, improvements to the landscape and the creation of industrial estates. A new shopping area called The Forge was built.

Later in the 1990s, further regeneration of the Gorbals area of Glasgow took place. This area suffered from some of the worst housing conditions in Europe. The housing built after the war was unfit for purpose, as the high rise flats were damp, left people isolated through broken lifts and high unemployment lead to a gang culture and high crime rate. The Crown Street Regeneration Project replaced the high-rise flats and tenement flats with low-density housing. It also included services, landscaping and the fostering of a community spirit.

The decline of the heavy industry and shipyards along the Clyde resulted in the development of new industry to generate income for the city. The SECC was built on old dockland at Anderson Quay, attracting visitors and business to the area. In recent years Pacific Quay has also been developed, housing attractions such as the Glasgow Science Centre and the IMAX cinema complex. The River Clyde is now multi-functional, being used for tourism, transport as well as industry. Modern apartments have also been built at Glasgow Harbour, catering for a variety of people from retired to students. Transport links over the River Clyde have been improved with the addition of the Clyde Arc (known as the Squinty Bridge).

Figure 2.6 Regeneration along the River Clyde

Rural/urban fringe developments

The rural/urban fringe is the boundary where the urban area meets the countryside. There is a mixture of land use in this area such as housing, industrial estates, golf courses and airports. The mixture of land use often causes conflict as different groups have different needs and interests.

Science parks, business parks and industrial estates locate in the rural/urban fringe as the land is cheaper, there is room for expansion and good transport links make for the easy and efficient export and import of goods.

Golf courses, leisure parks and other recreational land users locate here because of the large amount of land they take up and which is available at a reasonable price. Housing has grown here and small villages have expanded as more people move out of the cities and commute to work.

Out-of-town shopping centres want to locate in the rural/urban fringe, as they need lots of space as well as room for parking, as most people arrive by car. Additionally, good transport connections and cheap land encourage them to establish there.

Farming is a major use in the rural/urban fringe, although the farmers often come under great pressure to sell their land for development. And a farmer will make far more money from a sale if there is already planning permission for building to occur on the land.

Measures to resolve these problems

- Under '**greenbelt**' legislation passed in Britain in the 1950s, strategies involved invoking planning restrictions on developments such as housing, industry, landfill sites and recreational centres.
- Smaller towns and villages were identified for growth. Throughout Britain new towns and overspill areas were used as growth centres to limit further development within the rural/urban fringe.
- However, small rural villages remained popular with people wishing to leave the city and were still a target for developers. Limitations on the number and type of buildings have kept the rural environment protected from the worst excesses of city developers.
- The needs of industry and opportunity for offering employment had to be balanced with the desire to protect the rural/urban fringe. This has generally proved very successful.

Example 🚩

Look at Figures 2.7 and 2.8: Central Business Districts of Edinburgh and Glasgow. For a named city you have studied, describe in detail the recent changes which have taken place in the Central Business District.

Figure 2.7 Edinburgh city centre

Figure 2.8 Glasgow's Buchanan Galleries

Sample answer

The city I have studied is Glasgow. Many of the buildings are old as some were used as warehouses like Candleriggs, but they are no longer used so are converted into houses, clubs and hotels (✓). This makes use of valuable space as there is not a lot of empty land in the city centre to build on and it is expensive (✓). These converted buildings allow new people to buy flats in the city centre (✓). Some streets have become

⇨

⇨

pedestrianised to make it safer for shoppers (✓), for example, Buchanan Street (✓). This reduces the number of vehicles in the city centre (✓) and also reduces pollution (✓). Indoor shopping malls like Buchanan Galleries have been built to allow people to shop in all weathers as Glasgow gets lots of rain (✓).

Comments and marks

This is a very good answer. The question asks for a named city and this is stated at the beginning of the answer. It gains marks for showing knowledge of the case study by mentioning named examples like Buchanan Galleries. It covers changes in land use, transport and the environment. It mentions at least six points so gains the full **6 marks out of 6**.

Developing world cities

In cities in the developing world, changes to urban areas have involved developments related to squatter areas and shanty towns. These developments can be illustrated by referring to a particular city in the developing world, such as Sao Paulo in Brazil.

*Kibera
Nairobi Kenya*

Sao Paulo

Sao Paulo is sited on a plateau near the confluence of the Tiete and Tamanduate rivers. The city is 750 metres above sea level and separated from the coast by an escarpment called the Serra do Mar. The coastal location offered advantages for port development with associated industries such as warehousing, food processing, petrol chemicals and an integrated steelworks. The population increased due to factors such as high birth rates and decreasing death rates and migration into the city from rural areas.

This growth has led to a number of problems including the growth of squatter areas and shanty towns. Similar problems exist in cities throughout the developing world.

Typical problems

- Lack of housing results in homelessness. Families consequently have to live on the streets, in squatter areas or in shanty towns. Lack of proper facilities such as clean water, electricity and sewage disposal systems lead to health problems, including the spread of diseases such as typhoid and cholera. Inadequate water supplies are common and in the worst cases people use polluted wells and open pipes. Infant mortality rates often increase and life expectancy decreases as a city's population grows.
- Lack of industry and high demand for jobs affect employment opportunities. This creates problems of low income and poor standards of living, involving a lack of food, poor education and little healthcare.
- There are shanty towns where people live in makeshift housing without basics such as electricity, cooking facilities, proper sanitation

or access to clean water. Crime rates are often very high due to extreme poverty. Local police may be unable to deal with the level of crime due to lack of manpower.

Measures taken to solve problems in shanty towns and squatter areas

- Cities such as Sao Paulo that have been affected by these problems have benefited from the involvement of international aid schemes. These have provided money, medical help and educational services. Funding is often obtained from sources such as United Nations sub-agencies, for example the United Nations Education, Social and Cultural Organisation (UNESCO), the World Health Organisation (WHO) and the World Bank.

- City authorities have introduced primary healthcare schemes and local self-help schemes, particularly in shanty towns. These self-help schemes involve improving infrastructure in an area using local labour. Improvements to water supplies in many residential areas have been treated as a priority. Materials have been supplied to help improve the poorest quality houses. In some cities efforts have been made to re-house people in new town areas. Financial incentives have sometimes been offered to residents to move to other parts of the city. This has often been due to the expansion of commercial enterprise and the need for land, rather than simply efforts to help residents of shanty towns.

Key words and associated terms

Central Business District (CBD): The zone that contains the major shops, businesses, offices, restaurants, clubs and other entertainments in an urban area. It is normally located at the centre of the settlement at the junction of the main roads.

Commuter settlements: Small settlements on the outskirts of major towns and cities where people live who travel into the main settlement for employment and services.

Dereliction and urban decay: This refers to closed and abandoned buildings such as mines, offices or industries. They often become a source of visual pollution if they are not demolished.

Greenbelt: Area surrounding cities and towns in which laws control development such as housing and industry in order to protect the countryside.

Inner city: This is the area near the centre of a city that basically consists of the older manufacturing zone and the zone of low-cost housing.

Land use zones: These are areas of a settlement where certain functions are dominant, for example the industrial zone.

Pedestrianised zones: These are traffic free areas, usually within a city centre. People can walk and sit in the street because traffic is forbidden to enter.

Renewal and regeneration: This refers to the processes by which older areas are demolished and replaced by new buildings, which often have totally different functions from the original building or area.

Settlement function: Functions are individual activities that settlements perform, such as commercial, industrial, administrative, transport, religious, medical, recreational and residential.

Traffic congestion: This is a heavy build-up of traffic along major routes and within the city centre, which causes serious problems and pollution in many cities.

Change in rural areas in developed countries

Farming has changed greatly over the last 50 years. Although the main farming **processes** have stayed the same, the way they are carried out has changed significantly. Some processes include:

- preparing the land for crop growing, ploughing, seeding and harvesting
- tending livestock
- transporting produce to market
- maintenance of equipment
- using artificial **drainage** and irrigation
- using labour and machinery on the farm.

The changes have affected methods, organisation, farm output, labour, farming landscapes and the overall status of farming within the economy.

Modern developments in farming include new technology, diversification, government policy, organic farming and genetically modified crops.

Key point

You should know about change, problems and policies in rural areas in developed and developing countries.

The impact of new technology

There has been an increased use of machinery like tractors, ploughs and combine harvesters which has resulted in increased crop yields, as work is done faster and more efficiently. Crops are harvested quicker so arrive at market fresher resulting in a better profit for the farmer. The use of sprinklers systems (artificially watering the crops) allows crops to be grown in areas where seasonal droughts occur.

Farming methods have become increasingly more scientific, with advances in medical care for animals, new improved seeds, chemical **fertilisers** and **insecticides** being used on a much wider scale, as well as greater use of computers.

The use of the Global Positioning System (GPS) and Geographic Information Systems (GIS) has made precision farming possible. These technologies can be used for things like farm planning, field mapping, soil sampling and tractor guidance. This means a more precise application of pesticides, herbicides and fertilisers. Better control of the dispersion of such chemicals is possible through precision agriculture, thus reducing expenses, producing a higher yield and creating a more environmentally friendly farm. GPS allows farmers to work during low visibility in conditions such as rain, dust, and fog, saving time and increasing efficiency.

Increased use of machinery has meant that the farmer needs to spend less money on wages but this does increase the number of unemployed agricultural workers. This has led to rural depopulation as people leave the countryside looking for work elsewhere.

Machinery is expensive, however, and some farmers cannot afford to buy or repair items, so they either amalgamate with other farms creating huge farms, or they sell up causing smaller, family farms to disappear. Machinery is noisy and creates air pollution. Because large machines need a large area to work in, hedgerows are removed to accommodate them. This results in the loss of animal habitats. Another concern is that, without hedgerows, there is less protection from the wind so soil erosion can occur without the necessary roots to bind it together.

Chemical fertilisers and pesticides increase crop yield and allow crops to be grown on poorer lands, but they have an adverse effect on the environment as they can be washed out into rivers affecting the habitat and the water supply. There is also increasing evidence to suggest that certain pesticides have had a devastating impact on native bee populations. As key plant pollinators, such a drastic decline in their numbers is very troubling.

Diversification

Diversification is the generation of an income through non-farming activities. In order to reduce costs and increase the profitability of farms, farmers have increasingly added new non-farming land uses to their farms. All of this means that the farmer's income is not solely dependent on farming.

Much of this is linked to the leisure and tourist industry.
- Areas of farms are now used as golf courses or bike/rally tracks.
- Cottages once occupied by farm workers, now made redundant, have been converted into holiday homes for tourists.
- Areas have been set aside for camping and caravan sites.
- Some farmhouses offer bed and breakfast accommodation.

With the resulting increase in tourism to the countryside, this has led to an increased income for some farmers, creating a better standard of living. Some services like local shops and bus services, which were in danger of being lost, have remained. The local population can also use the new facilities, also improving their quality of life. Another advantage has been the employment created in the rural area, helping to reduce depopulation.

However, the increasing numbers of tourists using the countryside has an adverse effect. Most tourists arrive by car creating traffic congestion on the narrow roads. They park across farmers' gates, restricting access to fields and park on grass verges, causing damage to plants. They climb over walls causing damage which costs the farmers time and money to repair. Larger numbers of tourists produce more litter and noise pollution as well. The creation of caravan sites can look out of place and affect the natural scenic beauty of an area.

The effect of government policies on developments in rural areas
- In Europe, the European Union (EU) has had a major impact on agriculture. It set up the Common Agricultural Policy (CAP) to try to improve and secure the standard of living of European farmers.

A variety of policies, including **quotas**, set-aside policies, subsidies and other legislation, has led to many different developments. Some have been for the better from the point of view of farmers, others for the worse.

- Output has increased greatly but this has often led to huge surpluses of produce, resulting in the infamous butter mountains, apple mountains and wine lakes. Produce is often stored in barns in order to reduce supply and maintain prices.
- To deal with this problem, farmers received payments from the EU for setting aside land rather than growing crops. The cost of this has been met by taxpayers in member countries of the EU.
- The EU has also paid development grants to poorer agricultural areas, such as southern Italy, to help farmers modernise their farms.

Since the UK has voted to leave the EU these policies may change when this happens.

The effect of UK government legislation

The UK government also produces policies that affect farmers. The aim is to support and develop British farming, to encourage sustainable food production, protect the environment and improve standards of animal welfare.

Rural development programmes support a variety of schemes where funding is available for a wide range of purposes, including woodland-management equipment, creating on-farm reservoirs and using water more efficiently. Funding is also available for broadband, rural business support, on-farm food processing, arable and horticultural productivity.

Countryside Stewardship (CS) provides financial incentives for farmers/land owners to look after their environment through activities such as:

- conserving and restoring wildlife habitats
- flood-risk management
- woodland creation and management
- reducing widespread water pollution from agriculture
- maintaining the character of the countryside
- preserving features important to the history of the rural landscape
- encouraging educational access.

Organic farming

Organic farming means producing crops without the use of artificial chemicals like pesticides and fertilisers and with the highest standards of animal care. All organic farms and manufacturing companies are inspected at least once a year and the standards for organic food are laid down in European law. Organic farmers work with nature using natural techniques and by-products such as animal manure to grow crops. Animals are reared without the routine use of drugs and antibiotics, which are commonly used in non-organic, industrial farming. This means a safer environment that encourages insects and wildlife. There is little water pollution as there are no chemicals to run off into streams and rivers.

However, organic farming requires a greater effort and is therefore more demanding and time-consuming for farmers. Since crop production is on a much smaller scale, the amount of crop produced is less. Also, since the soil has not been improved by the use of chemical fertilisers output is lower. The cost of producing organic food is higher so the price paid by customers in the shops is also higher. Customers are paying for the special care organic farmers place on protecting the environment and improving animal welfare.

Genetically modified (GM) food

Genetically modified food is the term most commonly used to refer to crops and plants created for humans or animals to eat which have been modified in the laboratory to develop their qualities or improve the nutritional content. GM foods first appeared on the market in the early 1990s.

Many crops have been altered to make them resistant to insects and viruses, thereby helping farmers to increase their crop yield and profit. Some of these foods have also been enhanced to be highly nourishing. Others have been modified so that they have a greater shelf life, like tomatoes that taste better and last longer. As these GM crops are more disease resistant, farmers need to use fewer potentially harmful pesticides so there is less pollution to the environment.

However, many people are opposed to GM crops as they feel the process is against nature. In some countries GM crops are banned until further research takes place on the potential long-term effects to human/animal health as well as their effect on the environment.

Change in rural areas in developing countries

New technology

In developing countries, agriculture is the main source of employment and way of life for between 50–90 per cent of the population. Of this percentage, small farmers make the up the majority, up to 70–95 per cent of the farming population. Many of them are subsistence farmers, growing only enough food to feed themselves and their families.

India has made large steps towards becoming a developed country, but farming is still one of its largest industries with more than 50 per cent of the population earning a living from it. Traditionally, farms were small and farmed intensively using manual labour. The main crop grown was rice. There was little use of machinery as this was too expensive so water buffalo were used to plough the fields. Rivers were used as irrigation for the rice, which needs a large amount in order to grow. However, with the continued increase in India's population, there was a need to produce more food.

In the 1960s changes were introduced to try to produce more food. The introduction of these new farming practices was called the Green Revolution.

> **Key points !**
> * You should know about recent developments in farming in developing countries, including the changes they have caused in rural areas.
> * What impact have new technology and political policies had on rural changes in these areas?

This involved the introduction of machinery, land reform, the use of pesticides and fertilisers, better road and rail systems, and irrigation systems.

The Indian government introduced various schemes to improve farming during a programme of 5- and 7-year plans. This involved a range of measures, including:

- land reform schemes whereby small farms resulting from the land inheritance system were amalgamated into larger farms
- helping farmers to borrow money to improve their farms
- the introduction of new and improved 'miracle seeds' in order to increase yields
- using chemical fertilisers to improve soil fertility
- increasing mechanisation by using tractors, motorised ploughs and other farm machinery
- employing agricultural advisers and setting up various training schemes for farmers
- spraying insecticides onto crops to prevent them being eaten and destroyed by insects
- introducing modern irrigation methods to replace inefficient methods, such as the use of traditional inundation canals
- introducing legislation designed to increase the size of fields and allow the system to use large machinery and become more efficient.

Intermediate technology

Much of the new technology available is too expensive for many farmers so they depend on the use of '**intermediate technology**'. This is technology which is more appropriate for their needs and skills and which they can afford. For example, instead of expensive irrigation schemes they can build stone lines across fields to trap water. Other examples include building terraces on steeper land to reduce soil erosion and using solar energy to pump water from wells.

Biogas plants convert animal and human sewage into gas, which is used for heating and lighting homes. These are quite simple and cheap to build.

Greenhouse farming is also becoming increasingly popular, allowing farmers to harvest crops when the price is highest.

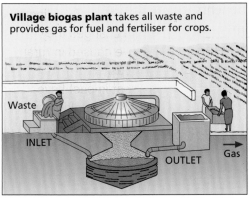
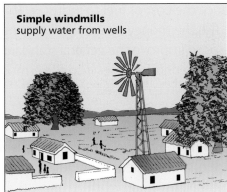

Figure 2.9 Examples of intermediate technology

Example 🚩

Describe in detail how recent developments in farming affect the people **and** the landscape in **developing** countries.

6 marks

Sample answer

The developments I am going to talk about are technology and biofuels. Pesticides reduce disease, producing better crops (✓), which in turn gives the farmer extra crops which he can use to trade (✓). Using fertilisers can give the farmer more crops making him a profit (✓), which can increase his standard of living (✓). The use of machinery means less strenuous work for the farmer (✓) so it is quicker and more efficient (✓). But there are bad effects as well. Using pesticides and fertilisers can affect people's health, as they get washed into the river which they use for drinking water (✓). It affects the environment as these chemicals can damage the fish (✓). There is an increased demand for biofuels, which means higher crop prices and can result in the farmer getting a higher income (✓) and create employment (✓).

Comments and marks

This is an excellent answer. The candidate identifies and gives examples of two new developments. The candidate talks specifically about these developments, giving lots of detail about the advantages of technology and biofuels. The answer then gives information on the disadvantages, gaining the remaining marks.

This answer scores **6 out of 6**

Biofuels

Biofuels are fuels produced from plant material and are usually made from crops grown specifically to produce biofuels. In Kerala, India, for example, the main biofuel crops are jatropha and sugar cane. Biofuels have both advantages and disadvantages.

Advantages

- Biofuels reduce the amount and cost of oil a country like India has to import.
- Using biofuels should reduce the cost of fuel to industry. This benefits local factories, which can use the money saved to employ more people, which, in turn, helps to improve the general standard of living.
- Burning biofuels puts fewer greenhouse gases into the atmosphere.
- Trees do not need to be cut down to be used as fuel.

Disadvantages

- If more land is used for growing the biofuel crops, there is less available to grow food. (Around one quarter of people in India do not have enough food.)
- Like other biofuel crops, sugar cane and jatropha use a lot of water, which is in short supply during the dry season. This places a further burden on local/regional water resources.

- Trees and forests have been cut down to make room for biofuel crops, meaning an increase in greenhouse gases and soil erosion.
- There is also evidence to suggest that the processes used to produce biofuels leave a considerable carbon footprint.
- They are expensive to make.

GM crops

High-yield variety (HYV) seeds were developed by scientists to improve food supplies and reduce famine in developing countries. These HYV or 'miracle' seeds can produce up to eight times more crops than regular seeds on the same area of land. There are both advantages and disadvantages of HYV seeds.

Seeds have been modified so they can grow on land that previously would have been unsuitable. They are shorter so can withstand poor weather such as monsoon rain and winds. The increased output means that more food is produced, feeding a greater number of people, thereby reducing starvation and famine. It increases farmers' profit so giving them a better standard of living.

Disadvantages include the taste – many people think the crops taste different. Using fertilisers and pesticides can harm the environment. They are water thirsty crops so irrigation is needed and this is expensive.

Key words and associated terms

Arable farming: Farms where the main activity and income source is the growing of crops.

Biofuels: Non-fossil fuels, obtained from various sources such as gas from animals and crops.

Cereal crops: Crops that are grain crops such as oats, barley or wheat.

Diversification: Adding different enterprises to a farm in order to improve income and allow the farmer to be less dependent on income from farm produce.

Drainage: If the underlying rock in an area is clay, bog and marshland may develop in the fields. Pipes are laid to drain excess water from the surface and allow the land to be farmed.

Fertilisers: Substances that are added to soil to increase fertility and improve crop yields. They may be organic or chemical.

High technology: This is the use of advanced, sophisticated machinery which requires a high degree of skill to operate.

Insecticides: Chemicals that are sprayed onto crops to kill insects that may be attacking the crops and damaging the yields.

Intermediate technology: This is machinery and equipment which is more advanced than basic, primitive equipment but not as much as high technology.

Low technology: This is equipment which is very basic and cheap, such as ox-drawn ploughs and wood-burning ovens.

Organic farming: producing crops and rearing livestock without the use of harmful chemical pesticides, herbicides and fertilisers, antibiotics and growth hormones and no genetically modified organisms or feed additives.

Processes: The wide variety of types of work done on a farm, including for example seeding, ploughing, harvesting and stock rearing.

Quotas: Limits imposed on farmers in order to limit the output of certain types of produce to avoid surpluses and therefore a drop in prices.

Section 3: Global issues

Candidates have to study two out of the following six global issues.

Chapter 3.1
Climate change

What is climate change?

Climate change is the process by which **climates** across the globe have been changing, particularly over the last 150 years, in terms of seasonal rainfall and temperature patterns. Much of this change is due to **global warming**, which in turn is due to an increase in the **greenhouse effect**. The greenhouse effect is a natural process in the Earth's atmosphere but it is increasing out of control mainly due to human contributions.

Global warming

Since the early twentieth century the mean surface temperature of the Earth has increased by about 0.8 °C, with about two-thirds of that increase occurring since 1980. Scientists predict that during the twenty-first century the global surface temperature will rise by a further 1.1 to 2.9 °C based on their lowest emissions estimate, or by between 2.4 and 6.6 °C for their highest emissions estimate.

What is the greenhouse effect?

The greenhouse effect is the process by which the Earth's atmosphere absorbs some of the Sun's energy, helping to keep the Earth warm. This effect has been intensifying due to the release of more and more so-called 'greenhouse gases', mainly carbon dioxide, into the atmosphere. These greenhouse gases come from various sources including industrial pollution, burning forests, car exhaust fumes and power stations. The resulting rise in the Earth's temperature is now beginning to affect the climate all over the world.

Causes of climate change

Physical reasons for climate change include:
- changes in the amount of solar energy given out by the Sun over time
- natural activities on the Earth, including volcanic eruptions and variations in the amount of atmospheric gases present

Key point

You need to know both the physical **and** human causes of climate change during the last 100–150 years.

- changes in the movement of the Earth in its orbit; slight shifts in the Earth's angle of tilt and its orbit pattern around the Sun have contributed to significant changes in the temperature pattern
- gases given off by rotting vegetation in tundra areas have affected global temperatures
- sunspots, which are flares which occur from time on the surface of the Sun, also cause an increase in temperatures on the Earth's surface.

Human reasons include the following:

- The widespread burning of **fossil fuels** and forested areas throughout the world has released various chemicals into the air, but particularly the main greenhouse gas, carbon dioxide (CO_2).
- Trees use up carbon dioxide in photosynthesis. The cutting down of huge areas of forest has also meant that there are fewer trees to remove carbon dioxide from the atmosphere.
- Increasing industrialisation has led to the release of air pollution from chimneys and factories. Air pollution is also produced by traffic, rubbish dumps and other similar sources. The **pollutants** released include carbon dioxide and nitrous oxide (NO_2), which are some of the main gases responsible for the greenhouse effect.
- Gases such as methane are released by large herds of livestock, particularly cattle. Methane is another powerful greenhouse gas.
- Release of CFCs (chlorofluorocarbons) from, for example, aerosols and refrigerants also causes increases in global temperatures.

Effects of global warming

The effects of global warming are expected to be strongest in the Arctic, where it is associated with the continuing retreat of glaciers, permafrost and sea ice. There has been a resultant rise in sea levels and changes to rainfall patterns, which in turn have led to the expansion of desert areas.

Increased temperatures are causing the ice caps to melt. As a result polar habitats are beginning to disappear, threatening species like polar bears and walruses. Melting ice causes sea levels to rise, which will threaten coastal areas and low-lying areas like the Fens in England and parts of the Netherlands that lie below sea level.

Global warming could also affect weather patterns, leading to more droughts in areas like Ethiopia or greater flooding in areas such as the UK. This can also lead to crop failures and problems with food supply. An increase in extreme weather conditions are likely, with more worldwide tropical storms and stronger hurricanes in areas such as Florida and Georgia along the east coast of the USA.

If temperatures increase, this could encourage tourism in some regions, but could also cause problems for others, such as the Swiss Alps, as there will be less snow in some mountain resorts. Global warming could threaten the progress of developing countries as restrictions on fossil fuel use may be imposed to slow the rate of increasing CO_2 levels.

In the UK, tropical diseases such as malaria could spread as temperatures rise. The livelihood of fishermen could be in danger if the sea temperature causes native species to move north to colder waters. Plant life would also be affected with some non-native species suddenly thriving in previously unsuitable areas on account of the higher temperatures, while native species begin to die back. Higher temperatures may cause water shortages in areas like the Mediterranean.

Managing climate change

- Responses to climate change and global warming have included efforts to reduce emissions of greenhouse gases through national and international agreements.
- Most countries have joined the United Nations Framework Convention on Climate Change (UNFCCC).
- A range of different policies have been adopted to try to reduce greenhouse gas emissions.
- Many developed and developing countries are aiming to use cleaner, less polluting technologies, such as **renewable energy**. There are also schemes to capture and store carbon released by the burning of fossil fuels.

Strategies

The Kyoto Protocol is an agreement reached in Kyoto, Japan, in 1997, committing industrialised nations to cut their greenhouse gas emissions. The Protocol was ratified by 192 parties. The USA did not and it dropped out in 2001. The signatories agreed to reduce their country's emissions to five per cent below 1990 levels between 2008 and 2012. The Kyoto Protocol took effect in February, 2005.

In December 2012 the Kyoto Protocol was extended to 2020 during a conference in Doha, Qatar. In 2015, a summit was held in Paris where the Paris Agreement replaced the Kyoto Protocol. The parties agreed to limit warming 'well below' 2 degrees, and below 1.5 degrees above pre-industrial levels, if possible.

- Other efforts to reduce carbon emissions have been blocked by countries that do not accept the scientific evidence of climate change and global warming.
- In Europe the argument is more about what action should be taken, whereas many in the USA still debate whether or not global warming is actually happening. The controversy surrounding climate change looks likely to continue worldwide for some time to come.
- Meanwhile, if agreement is not reached on how to control carbon emissions, climate change will continue and global temperatures will keep rising and their effects will be felt over large areas of the world.

Key words and associated terms

Climate: This refers to the average weather conditions in a particular area or region, usually measured from temperature and rainfall data taken over a period of 35 years.

Climate change: Changes to the average weather conditions over the last 150 years.

Fossil fuels: Fuels such as coal and oil obtained from the fossilised remains of plants and animals. When burned they produce gases that pollute the atmosphere, including carbon dioxide, the main greenhouse gas.

Global warming: A rise in average global surface temperatures. It began in the mid-nineteenth century and continues in the twenty-first century.

Greenhouse effect: The process by which the Earth's atmosphere traps energy from the Sun and warms the Earth. It is increasing due to the release of additional greenhouse gases into the atmosphere.

Pollutants: Materials that are released into the environment and ultimately cause damage to the physical landscape and atmosphere.

Renewable energy: The use of technology to produce cleaner sources of energy that can be continually renewed. Sources include hydroelectric schemes, wind power and wave power.

Key points

* The climate of any area is based on the average weather conditions recorded over a period of time varying from a year to several years.
* Two main features of the weather are usually recorded and shown on a graph, namely total monthly rainfall from January to December and average monthly temperatures throughout the year.
* For the purposes of the examination you need to know the climates of two main zones, **tundra** and equatorial.

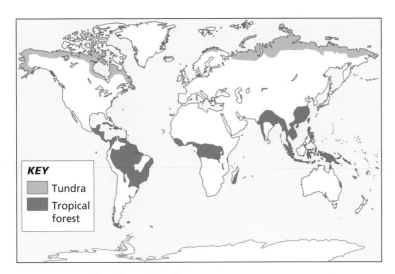

KEY
Tundra
Tropical forest

Figure 3.1 Map of tundra and rainforest areas

Climate graphs

In the examination you may be asked to identify and describe the two climates listed above. At National 5 you may also be asked to describe the climates in detail and perhaps comment on the impact of the climate on local people and environments.

- When describing climate graphs, make sure you refer to rainfall distribution throughout the year, for example the wettest and driest times. Quote actual amounts, such as the highest and lowest, and describe whether the climate is relatively wet or dry, noting exceptional conditions.
- Having described the rainfall pattern, describe the temperature pattern. Again, note the highest and lowest temperatures and the difference between these, which is the temperature range.
- Quote figures from the graph to give added detail. Describe the temperature pattern in general terms, noting whether the climate is cold, cool, mild, warm, hot or very hot. Use these terms to describe the various seasons of the year.

Tundra
Main features of the tundra climate

Key point !

You should be able to describe the tundra climate and ecosystem.

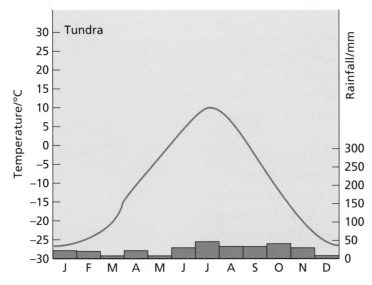

Figure 3.2 Tundra climate graph

The main features of this climate may be described as follows:
- the highest temperature is 10 °C in July
- the lowest temperature is −27 °C in January
- the range of temperature is 37 °C
- there is rainfall in every month
- the lowest rainfall is 8 mm in December, March and May
- total annual rainfall is low, less than 300 mm.

Main features of the ecosystem
- Tundra biomes are among the Earth's coldest and harshest.
- Tundra ecosystems are treeless regions found in Arctic areas and on the tops of mountains, where the climate is cold and rainfall is scant.
- Tundra lands are snow-covered for much of the year, until summer brings a burst of wildflowers. Hardy flora such as cushion plants survive on the mountain plains by growing in rock depressions where it is warmer and sheltered from the wind.
- The Arctic tundra supports a variety of animal species, including Arctic foxes, polar bears, grey wolves, caribou, snow geese and musk oxen. Mountain goats, sheep, marmots and birds live in mountain or alpine tundra and feed on the low-lying plants and insects.
- The summer growing season is just 50 to 60 days, during which the sun shines 24 hours a day.

- The few animals and plants living in these harsh conditions are essentially clinging to life. They are highly vulnerable to environmental pressures such as reduced snow cover and warmer temperatures caused by global warming. This global warming is seriously affecting the Arctic's **permafrost**, which is the foundation for much of the area's unique ecosystem.
- Permafrost is a layer of frozen soil and dead plants that extends for around 1500 feet (about 450 metres) under the surface.
- In summer in the southern regions the surface layer above the permafrost melts and forms bogs and shallow lakes.
- This leads to an explosion of animal life. Insects swarm around the bogs and millions of migrating birds come to feed on them.
- Due to global warming most of the permafrost is melting in the southern Arctic. Shrubs and spruce that previously could not take root on the permafrost now dot the landscape and this is altering the habitat of the native animals.
- The melting permafrost is also contributing to global warming. About 14 per cent of the Earth's carbon is contained in the permafrost.
- Until recently the tundra locked in huge amounts of carbon and kept it out of the atmosphere. As the permafrost melts, dead plant material releases CO_2 into the atmosphere. Instead of being a carbon sink, the tundra is now a carbon contributor.

Equatorial climate

Main features of the equatorial climate

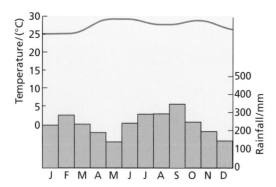

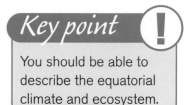

Key point

You should be able to describe the equatorial climate and ecosystem.

Figure 3.3 Equatorial climate graph

When describing this climate, refer to:
- annual temperatures, which are quite high
- the main characteristics are very little variation in the high temperatures throughout the year, with a maximum of around 32 °C and a minimum of 28 °C (giving a range of only 4 °C)
- rainfall throughout the year totals 2000 mm, with a monthly maximum of 340 mm and a minimum of 175 mm
- there is no seasonal variation in rainfall or temperature, except further north where conditions, especially rainfall, can vary.

Main features of the ecosystem

There are different layers of plants in the rainforests. Emergents are the tops of the tallest trees in the rainforest. These are much higher and so are able to get more light than the average trees in the forest canopy. Next are the tall trees that form a canopy. On the next level are vines, orchids and **epiphytes** that grow high up in trees to reach more sunlight. Lower down are tree ferns and similar short trees, and next there is the forest floor. Fan palms have large, fan-shaped leaves that are good for catching sunshine and water. The leaves are segmented, so excess water can drain away.

Rainforests have a shallow layer of fertile soil, so trees only need shallow roots to reach the nutrients. However, shallow roots can't support huge rainforest trees, so many tropical trees have developed huge buttress roots. They grow out from the base of the trunk, sometimes as high as 15 feet above the ground. These extended roots also increase the area over which nutrients can be absorbed from the soil. Competition at ground level for light and food has led to the evolution of plants which live on the branches of other plants, as well as aerial plants that gather nourishment from the air itself using so-called 'air roots'.

Animal life is plentiful. The rainforests teem with insects. Frogs and reptiles thrive. Bird life is noisy and colourful with many species of parrots and cockatoos. Since birds live mostly in the rainforest canopy, rainforest snakes are often tree snakes. Many mammals also live in trees, whether they are herbivores or carnivores, and rivers and waterholes contain different water animals.

A few examples of animal adaptations in the world's tropical rainforests include:
- camouflage
- the times at which they are active
- poison and other deterrents
- the interdependence on other species.

Many camouflaged rainforest animals, such as walking stick insects which look just like a tree branch and the slow-moving, green algae-covered sloths that hang from trees and blend in with their environment, use camouflage to avoid predators.

Many animals in the tropical rainforests of the world have adapted to either a night time or a daytime mode of life in order to survive.

Another adaptation is poison. The poison dart frog is famous for its bright colour, but in the animal world, bright flashy colours mean danger. The toxins and bright colours warn predators of the dangers of eating members of this frog family.

Use and misuse of these climatic areas

Tundra: Use and misuse

How does the tundra environment affect humans living there?

- Few crops or animals can be grown or reared due to the permafrost. Most food supplies have to be flown in, making them very expensive.
- In winter, snow storms often leave people in settlements cut off from the rest of the world. Travel and transport is made difficult throughout the year due to the ice in winter and the melting snow in summer. Wind chill can cause severe problems, such as frostbite and hypothermia, to people working outdoors.
- Heating costs are high because of the low temperatures, especially during the winter months. Buildings have to be raised off the ground on stilts to prevent them sinking during the summer months when temperatures are higher and the snow melts.
- In summer fish and seals are caught as the frozen seas melt.
- Oil is transported through the tundra in pipelines. These pipelines interfere with the migration routes of animals such as caribou.

Tundra: Use

The original inhabitants of the tundra area of North America are the Inuit (Eskimo). They have used the resources of the tundra for food through fishing and hunting seals. The tundra has not been used commercially by the native people. The activities of the Inuit have not damaged wildlife or the natural environment. Other developments include oil exploration, for example in Alaska.

Tundra: Misuse

Tundra areas have been misused in various ways which have damaged the natural environment and the habitats of local wildlife, usually for financial gain. Examples include:

- the commercial hunting of seals, especially young pups which have been slaughtered for their pelts
- oil spills and offshore accidents have caused pollution and contamination of surrounding seas
- the building of pipelines to transport oil from the oil ports to other parts of the country; these interfere with the migration routes of local wildlife such as the caribou
- heavy trucks bringing in supplies cause increased carbon emissions and can damage roads.

Strategies adopted to reduce the impact of misuse

These include:

- building the Alaskan oil pipeline on stilts to allow caribou ease of passage when migrating
- using ice roads with transport that can adapt to them; these confine the transport of heavy vehicles to certain routes and also help to restrict carbon emissions to certain areas, reducing the environmental impact throughout the tundra
- supporting national and international efforts to reduce carbon emissions, which add to global warming and the melting of permafrost and ice.

Deforestation (tropical rainforests)

Throughout equatorial areas of the world, thousands of square kilometres of trees have been destroyed in various ways and for a variety of reasons.

Methods of destruction include:

- destruction by fire, which may be the easiest and quickest way to destroy forests so the land can be used for other purposes, such as farming
- **logging** – trees are cut down by loggers, since hardwoods such as teak and mahogany can be sold for large sums of money abroad
- trees destroyed by valleys being flooded to create reservoirs for multi-purpose water schemes
- areas of forest being removed to create space for settlements
- areas of forest being cleared to create room for new transport routes.

Despite the damaging effects on the environment of **deforestation** (see below), forests continue to be destroyed in countries such as Brazil and the Congo, and parts of India and south-east Asia.

This is partially because deforestation can be very profitable for certain groups. People profit from deforestation in various ways:

- cutting down hardwood trees to be sold for timber

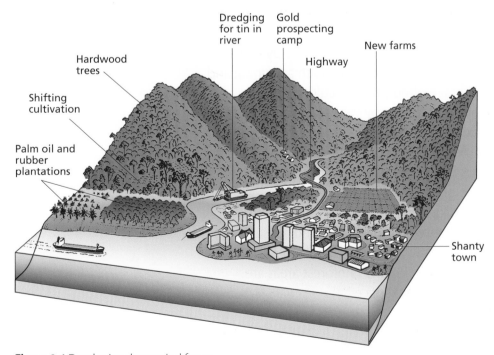

Figure 3.4 Developing the tropical forest

- clearing areas for farming
- clearing trees for mining of a variety of minerals, including iron ore, copper, bauxite and gold
- using trees for fuel such as charcoal in local industries like iron smelting
- clearing large areas of forest to create grazing land for **cattle ranching**.

Effects of deforestation on the environment and people who live in the forests

- Destruction of the forests can have a devastating effect on local people, on thousands of different species of plants, animals and insects, and even on the atmosphere.
- Deforestation is known to contribute to global warming and the greenhouse effect, leading to climate change and increases in sea levels throughout the world.
- Without trees to interrupt run-off of rainwater, the result can be serious flooding causing widespread disaster and thousands of deaths.
- Flooding due to deforestation can degrade soil and mining (a common reason for removal of forest) causes increased pollution.
- Rivers can become polluted, killing fish and affecting local tribespeople.

Strategies to reduce land degradation

- Strategies to manage and reduce deforestation include:
  - **reafforestation** with mixed trees
  - the use of crop rotation by farmers which helps to limit the amount of forest which is removed for farming
  - the purchase of forest areas by conservation groups
  - returning forests to native peoples.
- Several of these schemes have been very effective but outside interests in mining, ranching and so on often take precedence over conservation measures.
- There have been attempts to control deforestation through government legislation but the impact has been limited due to economic demands for development.
- Various governments have passed laws to protect forests by limiting the amount of land that can be used for activities such as mining and ranching. However, these laws are often very difficult to enforce.
- Worldwide, there are ongoing campaigns by various environmental groups such as **Greenpeace** to save the forests, for example by direct protests to specific governments and encouraging other commercial developments within forests by buying sustainable forest products such as tropical fruits, and so on.
- There are many who live in the forest who can work in harmony with their environment, including tribes who hunt, gather and are subsistence farmers, rubber tappers and loggers who use sustainable methods.
- Unfortunately, despite all these efforts many rainforest areas are continuing to be reduced at an alarming pace.

Example

Describe, in detail, humankind's use and misuse of either a rainforest or tundra area you have studied.

6 marks

Sample answer

There are many uses of the rainforest but there are also misuses. One use of the rainforest is tribes. The tribes grow their own food in the rainforest which helps them survive (✓). Another use of the rainforest is the plants used for medicine (✓) like the rosy periwinkle, which is used to cure childhood cancer (✓). The tribes use the plants for medicine for themselves which is another use of the rainforest (✓). Another use of the rainforest is that animals use the trees as their homes as it provides them with food and shelter (✓). A misuse of the rainforests is cattle ranching. People need to cut down the trees and use up a lot of land for the cattle when it could have been used for other things (✓). Another misuse of the rainforest is deforestation. People cut down trees, which can lead to animal habitats being destroyed and the spoil would then be washed away (✓). Another misuse of the rainforest is logging. People use the rainforest for logging when it could be used for something else.

Comments and marks

This is a good answer as it covers both uses and misuse. However, it is a bit repetitive in places, for example, a repeat is made of the use of medicine. Logging did not receive a mark because it has the same reason given as cattle ranching. Even so, it still achieved **6 marks out of 6**.

Key words and associated terms

Cattle ranching: Rearing of large herds of cattle on areas of cleared forest.

Deforestation: Removal of trees, usually on a large scale.

Degradation: The process of reducing land that was formerly productive to unproductive land.

Epiphyte: a plant that grows on another plant

Greenpeace: An international organisation that organises protests against the causes of world pollution and other elements that damage the natural environment.

Logging: A commercial business that cuts down trees to provide timber to sell.

Permafrost: A layer of permanent ice that lies just below the surface in tundra areas.

Reafforestation: The process of replanting trees in former forested areas.

Tundra: A region that lies between the polar region of perpetual snow and ice and the northern limit of tree growth in the northern hemisphere.

Earthquakes

- Earthquakes occur when rocks deep within the Earth's **crust** move suddenly. This movement causes shockwaves to travel outwards in different directions through the crust. The source of the shockwaves is called the focus and the point immediately above the focus is known as the **epicentre**. This is where the most severe shockwaves occur.
- There are three main types of shockwave:
  - Push or primary waves ('P' waves) cause rocks to move up and down and are the fastest type.
  - Shake or secondary waves ('S' waves) cause sideways movement of rocks and move at about two-thirds the speed of 'P' waves.
  - The third group are known as long waves ('L' waves). They move along the Earth's surface and, although they are the slowest, they are the most destructive.
- The strength of these waves can be measured using an instrument called a seismometer and is recorded on a seismograph. The scale used to describe an earthquake's strength is called the Richter Scale.
- The distribution of earthquakes is very similar to that of volcanoes (see Figure 3.5). Both occur along or near the boundaries of the large **plates** that make up the Earth's crust. These areas are called the plate margins. The Earth's crust is weakest at the margins of the plates. The plates are 'floating' on top of the Earth's **mantle** layer and are moving very slowly relative to one another.
- There are three types of plate boundaries, namely constructive, destructive and sliding. At these boundaries the plates are moving relative to one another in different directions.

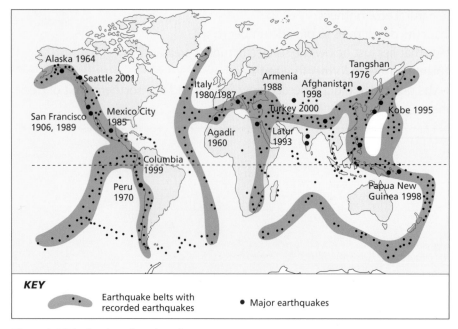

Figure 3.5 Distribution of earthquakes

Constructive boundaries

In these areas the plates are forced in opposite directions, causing rocks in these areas to be put under a great deal of tension. Eventually the rocks break and move sharply, causing shock waves to travel through to the Earth's surface. These waves cause the ground to shake, creating an earthquake.

Constructive
plate margin

Destructive boundaries

In these areas one of the plates is being forced down beneath the other. A large amount of friction is created due to this and this friction stops the plates from moving. As the pressure continues to increase, the crust eventually moves suddenly downwards into the mantle, causing shockwaves that create earthquakes on the surface.

Destructive
plate margin

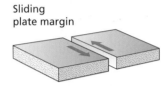

Sliding boundaries

In some areas crustal plates slide past each other. This sliding movement causes immense friction between the plates. As the pressure continues to build over time, it overcomes the friction and suddenly one plate jerks quickly past the other. This again causes shockwaves that in turn create earthquakes on the surface.

Sliding
plate margin

Figure 3.6 Different types of plate boundary

Volcanoes

- Most volcanoes are found near the boundaries of the crustal plates as shown in Figure 3.7.
- The most active volcanoes are located through the Mediterranean area, along the edge of the Pacific Ocean and in the middle of the Atlantic Ocean.

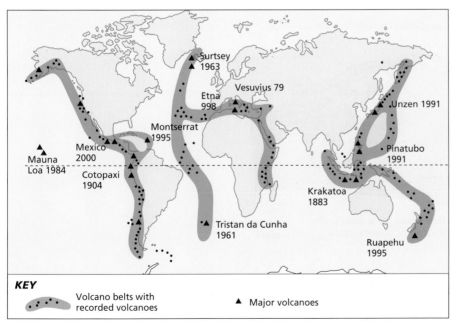

Figure 3.7 Distribution of volcanoes

Volcanoes may be described as **active**, **dormant** or **extinct**.

- Active volcanoes are those that are likely to erupt.
- Dormant volcanoes may not have erupted for some time but could again in the future.
- Extinct volcanoes are those that will never erupt again.

Causes of volcanoes

- Figure 3.8 summarises the main causes of volcanoes.
- Volcanoes occur where **magma**, ash, gas and water from beneath the Earth's surface are allowed to erupt onto the land and sea bed due to a weakness in the Earth's crust.
- These weaknesses are most likely to occur at the plate margins, especially at destructive and constructive margins or in areas where the plate is particularly thin. These areas are known as 'hot spots'.
- Pressure that builds up over long periods of time at the plate margins may be finally released by a volcanic eruption in which liquid magma is forced up through joints and weaknesses in the crust.

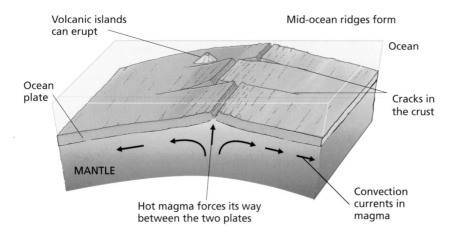

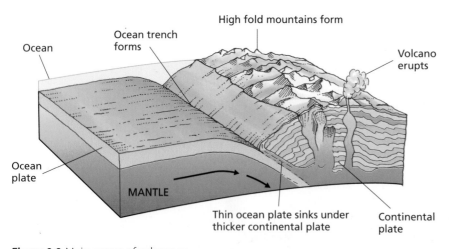

Figure 3.8 Main causes of volcanoes

Impacts on people and the environment

- For either an earthquake or a volcano that you have studied in class, such as the Mexico City earthquake in 2017, you should be able to discuss the underlying causes of these hazards in some detail.
- You should know the type of plate margin at which the hazard occurred, where the focus and epicentre was (for an earthquake), the strength of the earthquake, and the extent of the area affected by either the earthquake or the volcano.
- For the particular earthquake or volcanic eruption studied, you should refer to the damage caused to the landscape through earth movements (for earthquakes) or by **lava**, ash bombs, ash deposits, mudflows and landslides (for volcanoes).
- This could include damaged buildings, roads and infrastructure, and secondary impacts such as fire and flooding, people made homeless, injuries and deaths. You could also refer to the cost of the damage.

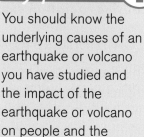

Key point !

You should know the underlying causes of an earthquake or volcano you have studied and the impact of the earthquake or volcano on people and the environment.

Mexico City 2017 – case study

On Tuesday 19 September 2017 a magnitude 7.1 quake – the deadliest to hit the nation since 1985 – struck shortly after 1 p.m. local time. The earthquake was centred south-east of Mexico City near Atencingo in Puebla state, about 120 km from the capital. It caused violent, prolonged shaking, which flattened buildings and sent masonry tumbling onto streets, crushing cars and people in the capital, Mexico City, and surrounding areas. Around 270 people died with 26 people – 21 children and five adults – being killed when the Enrique Rébsamen elementary school collapsed in Mexico City's southern Coapa district.

The earthquake toppled dozens of buildings, broke gas mains and sparked fires across the city and other towns in central Mexico. Much of the region was plunged into darkness, with 40 per cent of Mexico City and 60 per cent of the state of Morelos still without electricity early on Wednesday. Thousands of residents fled onto the streets. An earthquake two weeks previously was more than 30 times more energetic than the latest one, but despite this, the 19 September quake caused greater damage as it was closer to more populated areas and in Mexico City.

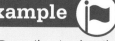
Example ⚑

Describe, in detail, the effects of an earthquake or volcano on the people and environment of the area. You should refer to an area you have studied. **6 marks**

Sample answer

The earthquake I studied was the Japanese Earthquake of 2011. It also produced a tsunami. This earthquake and tsunami killed around 19,000 people, it was disastrous (✓). Thousands of people were injured because of it and it was so destructive that over 500,000 people were made homeless (✓). Power and food shortages occurred for several days, which was a real struggle for them. They were also exposed to high levels of radiation. As for the impact on the environment, buildings collapsed and roads buckled ⇨

affecting transport in the area (✓) as well as cars being buried in the rubble (✓). Bridges had collapsed and due to its high magnitude several oil refineries went on fire causing a fuel shortage (✓). The nuclear power plant at Fukushima Daiichi went into meltdown causing radiation exposure (✓).

Comments and marks

This is a very good answer. The candidate has identified a case study then made reference to specific impacts on Japan. The question asks for impacts but saying it was a real struggle for them is not enough for a mark. The answer must refer to an impact. However, the candidate mentions both the people and the environment and refers to a case study so achieves **6 marks out of 6**.

Tropical storms

Distribution

Tropical storms are essentially very deep depressions with wind speeds varying from 60 km/hr to over 200 km/hr. They form over oceans within 30 degrees of the equator and begin on the eastern side of the oceans, moving westwards before dying out over land. When tropical storms reach speeds of 120 km/hr plus, they are described as **hurricanes**.

Figure 3.9 shows the distribution of tropical storms and Figure 3.10 illustrates the main features of this hazard.

Key point

You should have a detailed knowledge of a tropical storm and you should know and understand their distribution and general causes.

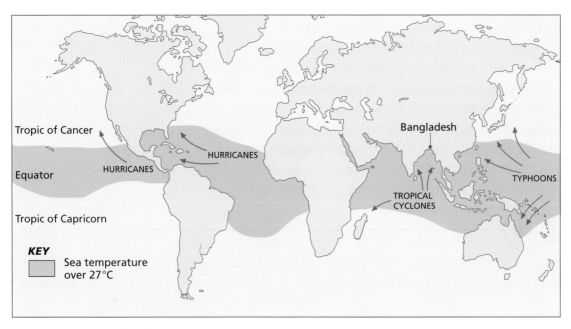

Figure 3.9 Distribution of tropical storms

Main features and causes

- Tropical storms form over the ocean in the tropics during the summer. High temperatures cause strong evaporation in deep ocean areas where the hot surface water can reach temperatures over 27 °C.
- Warm, moist air rises rapidly, cools and condenses forming very deep cumulonimbus clouds with heavy rainfall.
- Low pressure develops and as air is sucked into the depression, high winds result, increasing from gentle speeds to speeds of at least 60 km/hr.
- The rotation of the Earth encourages violent winds to rotate around the central '**eye**' of the storm. These winds are anticlockwise in the northern hemisphere.
- In the central eye, the pressure is very low with descending winds, increasing temperatures, clear skies and calm, dry conditions.
- The whole storm moves forward very quickly, bringing torrential rain, very high winds, falling temperatures and huge clouds in the areas immediately before and after the eye.
- Coastal areas over which the storm passes may experience storm surges due to the low pressure, causing the sea level to rise with huge waves forming in shallow coastal waters.
- Low-lying coastal areas are particularly vulnerable to massive damage and loss of life.
- As the storm moves away, the pressure rises, winds decrease, temperatures rise and the heavy rain turns to showers.

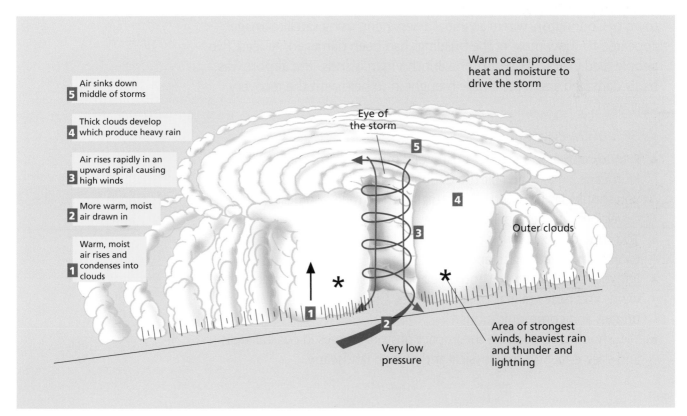

Figure 3.10 Features of a tropical storm

The impact of tropical storms

Key point !

You should be able to discuss the impact of a tropical storm on people and on the landscape.

Hurricane Irma 2017 – case study

Hurricane Irma devastated the Caribbean Islands and left a path of destruction all over the state of Florida. At least 72 people died. The most powerful Atlantic storm in a decade left a trail of destruction in the Caribbean, affecting an estimated 1.2 million people. Irma reached category five status, the highest, and packed sustained winds of up to 295 km/h (185 mph).

The storm cut a devastating trail across Caribbean countries and territories, killing at least 37 people there. It then moved up through the US states of Florida, Georgia and South Carolina, where an estimated 12 people died, before weakening into a tropical depression. However, cities like Miami and Jacksonville suffered flooding, and some 60 per cent of homes across Florida were left without power. The storm also brought torrential rain to Georgia, South Carolina and Alabama.

Hurricane Irma left the British overseas territory of Barbuda 'barely habitable'. Most of the island's population of just over 1,600 live in the town of Codrington, where an initial assessment, using satellite images, appeared to show most of the buildings had been damaged. At least five people died in Tortola, part of the British Virgin Islands. The airport was badly damaged and troops were brought in to help with the recovery operation.

When answering a question:

- If, for example, you refer to a particular tropical storm such as Hurricane Irma, which hit the Caribbean and the USA in 2017, you should say how and where this storm formed and describe its development over a period of time.
- You should also describe the strength of the storm and the path it took from beginning to end.
- For your study area you should be able to describe the effects of the winds and heavy rainfall on local forests, agricultural areas, buildings, bridges and communications.
- You should also describe the impact of storm surges in coastal areas and their effect on the physical and human landscape.

Example

Study Figure 3.11.

Describe in detail the distribution of tropical storms.

4 marks

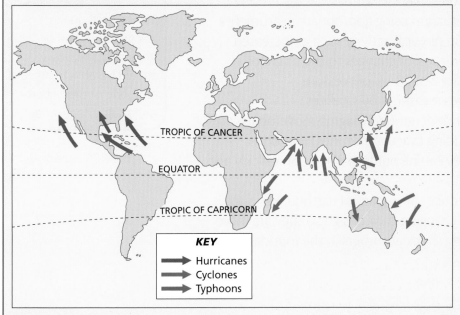

Figure 3.11 Tropical storms

Sample answer

Tropical storms are found close to the tropics of Cancer and Capricorn (✓), mainly off the southern coast of North America and the coasts of Australia, South Africa, China and India (✓). This is due to the high temperatures of the water found in these areas (over 27 degrees C) and the depth of this hot water (at least 6 metres), which causes the water to evaporate and rise in the atmosphere where it condenses and starts to spiral due to the spinning of the Earth. This creates a low pressure so other air is sucked in over the ocean to replace it, creating strong winds. The process occurs until the storm reaches land where it loses energy as it has no more warm air to fuel it, so eventually it just disappears.

Comments and marks

This answer makes two valid descriptive points in the first sentence. The rest of the answer is explanation and, although correct, it does not answer the question as only description from the map is wanted. The other two marks could be gained by describing the directions or naming other areas where the storms start or move to. This answer would gain **2 marks out of 4**.

Management of environmental hazards

Earthquakes and volcanoes

Prediction and planning measures may include:

- careful monitoring of the situation using appropriate instruments
- watching out for increases in physical activity such as increased earthquake activity, gas emissions, or changes to the shape of volcanoes that could indicate a build-up of pressure
- ground temperature changes measured by satellites
- use of ultrasound to detect underground magma movements
- planning evacuations (including evacuation zones) and other emergency measures so they can be put into operation quickly and effectively if necessary
- introducing measures to reduce the impact of earthquakes, such as strengthening buildings, changing their design so they are more 'earthquake-proof', and using shock absorbers in the foundations and structures of buildings
- flexible gas, water and power lines
- education programmes
- a good communication system to alert local people of impending disaster in plenty of time.

Key point

For an earthquake, volcanic eruption or tropical storm, you should be able to discuss methods of prediction and their effectiveness.

Tropical storms

You should be able to discuss methods of prediction and planning such as:

- monitoring and tracking storms using satellites and aircraft
- issuing of storm warnings in good time
- putting in place evacuation plans and procedures
- encouraging personal preparation by local people, such as building storm shelters, storing food and water supplies and boarding up windows
- preparation of emergency services.

In developing countries additional measures may include:

- building up and strengthening of river banks and coastlines; this is because many people in developing countries, for example Bangladesh, live on the low-lying banks of rivers and coastal areas
- programmes to educate people against risk
- aid programmes from developed countries to provide additional protection.

Responding to environmental hazards

Strategies adopted in response to environmental hazards include:

- short-term aid donated by non-governmental organisations (charities), governments around the world and international organisations such as the United Nations to provide emergency relief such as search and rescue teams, medical supplies, food, tents and so on
- long-term aid, again given by a variety of organisations, governments and individuals, to help the region and people recover and develop

(an example of this would be improving the infrastructure so buildings and roads are better able to withstand another hazard in the future)

- insurance pay-outs that help to pay for damage to businesses and homes; however, this only applies to those businesses and individuals that are insured and many of those in less economically developed countries are not
- reviewing and improving the planning and prevention measures outlined above in case of another hazard.

Key words and associated terms

Active volcano: Volcano that has erupted recently.

Crust: The outer layer of the Earth.

Dormant volcano: Volcano that has not erupted for a long time.

Earthquake: A vibration of the Earth's crust caused by shockwaves from sudden movements within the crust.

Epicentre: The point on the Earth's surface that is immediately above the source of the shockwaves.

Extinct volcano: Volcano that no longer shows signs of volcanic activity.

Eye (of the storm): An area of very low pressure found at the centre of a tropical storm.

Hurricane: Another name for a tropical storm.

Lava: Molten rock that pours out of a volcano during an eruption.

Magma: Molten rock that lies beneath the Earth's crust.

Mantle: The layer that lies immediately below the Earth's crust.

Plates: The Earth's crust is divided into seven large and twelve small sections known as plates. Plates are either continental or oceanic.

Tropical storms: Very deep depressions with strong winds and heavy rain. In different parts of the world they are called hurricanes, cyclones and typhoons.

Chapter 3.4
Trade and globalisation

- Some countries have a wide range of natural resources such as minerals, good water supply and a favourable climate, combined with high industrial and agricultural output and a highly developed level of technology. Such countries include the USA, Japan and the main countries of Western Europe such as the UK, France and Germany. These countries tend to dominate world affairs and have great influence economically and politically throughout the world.
- Often these countries act together as political or trading partners in alliances such as the North American Free Trade Agreement or the European Union. As a result they are in a position to dictate favourable terms to other countries, particularly those that are less developed. Often international companies from the dominant countries exploit the resources of the less developed countries and have used these to increase wealth and prosperity in the richer parts of the world.

Regions of the world are linked through trade

Key points !

* Trade is the exchange of goods and services between one country and another or between groups of countries.
* Links between developed countries and developing countries are often established through trade.

Table 3.1 shows the world's top 20 ranked countries (and the European Union) in terms of the total amount of trade in billions of US dollars for the year 2011. It is interesting to note that Europe has the highest amount of trade.

Table 3.1 Global exports table 2016

Rank	Country	Total exports (USD)	Date of information
1	China	$1,990,000,000,000	2016 est.
2	European Union	$1,900,000,000,000	2015 est.
3	USA	$1,456,000,000,000	2016 est.
4	Germany	$1,322,000,000,000	2016 est.
5	Japan	$634,900,000,000	2016 est.
6	South Korea	$511,800,000,000	2016 est.
7	France	$507,000,000,000	2016 est.
8	Hong Kong	$502,500,000,000	2016 est.
9	Netherlands	$495,400,000,000	2016 est.
10	Italy	$454,100,000,000	2016 est.

source: adapted from source www.cia.com

⇨

78

Table 3.1 (cont.) Global exports table 2016

Rank	Country	Total exports (USD)	Date of information
11	United Kingdom	$407,300,000,000	2016 est.
12	Canada	$393,500,000,000	2016 est.
13	Mexico	$374,300,000,000	2016 est.
14	Singapore	$361,600,000,000	2016 est.
15	Switzerland	$318,100,000,000	2016 est.
16	Taiwan	$310,400,000,000	2016 est.
17	United Arab Emirates	$298,600,000,000	2016 est.
18	Russia	$281,900,000,000	2016 est.
19	Spain	$280,500,000,000	2016 est.
20	Belgium	$277,700,000,000	2016 est.

You should be able to look at trade figures and describe the 'pattern' of trade between countries.

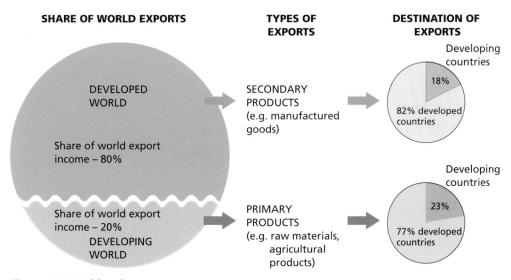

Figure 3.12 World trade

Patterns of trade

- Types of **exports** and **imports** include raw materials, finished and semi-finished products. Payments for services such as insurance payments, interest on loans, transport services and legal services provided by one country for another are another form of trade.
- This becomes important when thinking of '**balance of trade** situations'. Balance of trade is the difference between exports and imports. Countries generally do well economically if they export more than they import. They do not do so well if they import more than they export, because they are basically paying out more than they are earning.

Trade throughout the world is not always fair and equal:
the poorer countries of the world have been providing the wealthier countries with resources such as oil, timber, agricultural and mining products, often at very low prices; these materials are used in **manufacturing** industries and made into finished or semi-finished

products, which are then sold on at a much higher price, often back to the poorer countries; a good example of this is coffee. Coffee growers receive a much lower income and therefore lower profits from just selling the raw material instead of the finished, packaged product.

Many developing countries become dependent on this uneven trade with the developed countries and often end up in debt to them.

It is not easy for the poorer countries to change their situation by, for example, building factories to manufacture their own products.

- The developing countries may be locked into trading agreements with **multi-national companies** that favour the richer countries and to some extent exploit the developing countries.
- The developing countries cannot afford to withdraw from these agreements, since they would lose the vital income that they receive from exporting their raw material resources.
- Close inspection of trade figures not only reveals the pattern of trade but also gives a clue as to whether the trade pattern is fair and equal.

Strategies to reduce inequalities in world trade

Some developed countries of the world, with the support of international organisations such as the United Nations, have tried to introduce strategies to help reduce the inequalities referred to in the above section and the problems they cause. These measures have included the following.

Trade alliances

Some exam questions may ask about the advantages to a country of being a member of an alliance such as the European Union.

Advantages may include:
- having a wider market in which to trade
- trading with other countries on more favourable terms
- avoiding paying tariffs and import taxes (**barriers** to trade)
- being able to trade as a large group with countries outside the alliance
- being able to obtain grants and subsidies from the alliance to help poorer areas within member countries to develop
- the possibility of developing political and defence links with other member states
- freedom of movement of labour, i.e. people not having to apply for a permit to work in member countries
- a common currency such as the 'euro', so currencies do not have to constantly change from one country to another.

Disadvantages may include:
- loss of historical identity through, for example, giving up passports
- severing historical links with other trading partners throughout the world, for example Britain and New Zealand
- having to submit to the general laws and regulations of the alliance
- being unable to trade on favourable terms on an individual basis
- competition for jobs due to freedom of movement of labour

- having to pay higher taxes in order to provide funds to pay for the administration and various grants issued by the alliance
- having to subsidise economically weaker members
- possibly increased prices due to agreements between member states
- loss of political power
- possibly subsidising inefficient industry and agriculture in other countries
- loss of own individual currency.

Fair Trade

Other strategies that have been introduced include 'Fair Trade' agreements. This strategy attempts to 'cut out the middle man' by selling goods produced in developing countries directly to consumers in developed countries. This reduces inequality by ensuring that profits go back to the developing world **producers**, thereby helping them to obtain a regular income from fairer prices. Much of this income can then be reinvested or used to improve living standards in the developing country from which the goods originated.

Sustainable trade

Sustainable trade is when the commercial exchange of goods and services is beneficial to the social, economic and environmental structures of a country over both the short and long terms. Emphasis is put on concerns for the reduction of poverty and inequality, and the preservation and reuse of environmental resources. There is also concern for local traditions and working practices and great effort is made to ensure that the natural and human environments are not damaged. Fair Trade and organic trade are examples of sustainable trade.

Example 🚩

Fair trade producers

The fair trade system currently works with over 1.65 million farmers and workers.

There are 1,226 fair trade producer organisations across 74 countries.

1.5 hectares is the average size of the plot cultivated by a fair-trade farmer.

Look at the above facts concerning fair trade.

Explain the advantages that fair trade can bring to people in the developing world. **6 marks**

Sample answer

In fair trade the farmer gets more money as it cuts out the middleman who takes some of his profits (✓). If the farmer has more money then his family is better off, so they can improve their living conditions (✓). The money can be used to provide electricity or drinking water or pay for an education for their children (✓). Some fair-trade products are coffee, chocolate and bananas. We buy lots of fare trade products in the UK.

⇒

Comments and marks

This answer starts of well with the first three sentences gaining three marks. These sentences give three good reasons why fair trade is good for people in the developing world. However, the rest of the answer scores no marks as the points do not refer to the developing world and are simple descriptions. This answer scores **3 marks out of 6**.

Key words and associated terms

Balance of trade: The difference between a country's exports and imports.

Barriers: These consist of tariffs or taxes imposed on imported goods in order to protect home producers.

Exports: Goods and services sold to another country.

Imports: Goods and services bought from another country.

Manufactured products: Goods made from raw or semi-finished materials, for example cars, clothing and machinery.

Multi-national companies: These are large companies with branches operating in many different countries throughout the world, for example BP, Shell, Unilever and Coca-Cola.

Producers: These are companies or countries that manufacture or supply goods or raw materials. For example, Kenya is a producer of tea and BP produces petroleum and other oil-based products.

Tourism

Key point **!**

You should know about the main features of mass tourism and ecotourism.

- Tourism has become a global leisure activity. In 2016 there were over 100,184 million tourist arrivals worldwide. International tourism receipts grew to 1.22 trillion US dollars.
- The tourist industry is important and in some cases vital to the economies of both developed and developing countries on every continent.
- It creates massive opportunities for employment in the service industries. These industries include transportation services such as air transport, cruise ships and taxis, and hospitality services such as hotels, resorts, entertainment venues, amusement parks, shopping malls and theatres.

Tables 3.2 and 3.3 illustrate the leading countries for tourist arrivals and expenditure in 2015.

Table 3.2 The leading countries for tourist arrivals in 2015

International tourist arrivals 2015		
Rank	Country	(million)
1	France	84.5
2	USA	77.5
3	Spain	68.2
4	China	56.9
5	Italy	50.7
6	Turkey	39.5
7	Germany	35.0
8	United Kingdom	34.4
9	Mexico	32.1
10	Russian Federation	31.3

Source: World Tourism Organization (UNWTO)

Table 3.3 The countries who make the most from tourism

International tourism receipts (expenditure) 2015		
Rank	Country	US$ (billion)
1	USA	204.5
2	China	114.1
3	Spain	56.5
4	France	45.9
5	United Kingdom	45.5
6	Thailand	44.6
7	Italy	39.4
8	Germany	36.9
9	Hong Kong (China)	36.2
10	Macao (China)	31.3

Source: World Tourism Organization (UNWTO) July 2016

Mass tourism

Mass tourism refers to the global tourist industry involving hundreds of thousands of tourists visiting both developed and developing countries of the world. It is a form of tourism that involves tens of thousands of people going to the same resort often at the same time of year.

Growth of mass tourism

Mass tourism is dependent on developments in technology that have allowed the transport of large numbers of people throughout the world. Travel has become easier and cheaper, making it possible to travel further in less time and making places much more accessible. There is also a wide variety of airlines operating now, so competition for passengers makes flights cheaper and more affordable to a greater number of people.

Tourist resorts remain open throughout the year in summer and winter, allowing more people to visit. There is a wide range of holidays available and people generally have greater disposable income, so can afford to go on holiday. The Internet and holiday programmes on TV promote new and different places and activities making people more aware of the options available to them.

Additionally, people now receive more paid holiday allowance (annual leave) from their places of work. For example, in the UK the number of weeks' annual leave has increased from approximately two weeks in the 1950s to between four and six weeks today.

Impact of mass tourism

Advantages

In the Spanish town of Benidorm, mass tourism provides a wide variety of employment for the locals, for example, hotels, car hire, restaurants etc. As unemployment decreases, there is an improvement in the standard of the living. An increased demand for food provides farmers with a larger market for their produce, increasing their profit. New facilities like water parks are built for tourists but also benefit the locals. Tourist money can be used to improve the infrastructure of the region, for example, transport, water supplies and sewage.

Disadvantages:

Employment can be seasonal. Traditional ways of life can be lost as young people leave the countryside to live and work in the tourist resorts rather than on family farms, for example. Tourism creates pollution such as litter on beaches. Increased traffic causes noise and air pollution as well as traffic congestion in local villages. Large numbers of tourists visiting natural features like limestone caves can damage delicate structures. Beaches are eroded as sand is carried away on tourists' feet. Large areas of natural grassland/forest are removed to make way for new hotels etc., destroying the natural habitat of plants and animals.

Strategies to manage mass tourism

In Benidorm, the government limited the amount of development allowed along the coast with tighter control over the type and height of buildings. More sewage plants were built to prevent waste being dumped directly into the sea. Laws were passed to prohibit the dropping of litter and are enforced with fines. Bylaws were passed ensuring live music stops at midnight. In some hotels, all the room lights automatically switch off when guests leave, saving energy. Street lighting is low energy and many of the taps are foot pump-operated to save water. There are recycling bins everywhere in the resort to reduce litter in the resorts. Much of the food for hotels is sourced locally, ensuring farmers' livelihood and keeps the traditions of the local area. Improvements have been made to the local beaches and these are working towards the EU Blue Flag award.

Ecotourism

Ecotourism has grown extensively in many developing countries. Its goal is to avoid the social and economic problems associated with mass tourism. It also aims to increase awareness of the ecological damage that tourist development often entails. Developments have included setting up wildlife and safari parks in many African countries. Ecotourists are encouraged to visit these areas to gain an understanding of local culture and the lifestyles of the local population, along with an appreciation of the natural environment, wildlife and local ecosystems.

Impacts of ecotourism

Advantages

Ecotourism builds awareness and respect for the local culture, communities and environment. It employs local people, for example, in the making and selling of local crafts, in shops or as tour guides. The money from tourism improves the standard of living of the local area. Some of the money raised is used for conservation and sustaining the culture of the area, for example the Maasai Wilderness Conservation Trust which works to protect ecosystems and biodiversity, as well as wildlife monitoring through its community-based conservation. Ecotourism attempts to minimise the negative impact of tourism on an area and reduce the carbon footprint created by tourism.

Disadvantages

However, despite its many plusses, ecotourism jobs for locals often do not pay very well. Profits frequently go to other nations, as wealthy investors from these countries gain from the success of the project. Increasing tourist numbers potentially threaten **National Parks** and wilderness areas as more land is needed for tourist facilities, as well as the pollution the influx of people creates.

Some local residents may even be displaced as they cannot afford to stay or are forced to leave because of the development. These countries can become very dependent on money brought in by tourism, making their economy very vulnerable to market changes.

Strategies to manage ecotourism

In the rainforest, a tremendous amount of planning and organisation is needed to attract enough tourists to make money and still maintain the unspoiled forest and indigenous communities within them. Opening up an area to tourism without forethought can quickly destroy the forests on which the tourism is based.

On the Galápagos Islands, planners work out how many visitors each site can sustain. Only a certain number of visitors are allowed at a site at one time and the National Park Authority can reduce the number of tourists who are able to visit a site if they think it is being damaged.

Boat routes and tours are carefully managed so that areas are not overrun by visitors, reducing the amount of damage that might be caused. The introduction of entrance fees for visitors to the National Park also funds conservation projects within the Park. This allows tourists to enjoy the park while simultaneously ensuring the improvement of the lives of the local people, as well as protecting the local wildlife. Boat licenses have also been introduced, which help the Marine Reserve to police the waters far more carefully.

Example

Figure 3.13 Benidorm in 1936

Figure 3.14 Benidorm in 2014

Look at the images above.

Explain the impact that mass tourism has on the people and environment of an area.

You should refer to an area you have studied in your answer.

6 marks

Sample answer

An area I have studied is Benidorm. The tourists provide a lot of jobs for the local people in hotels and bars (✓). It reduces the number of people without jobs and gives them a better standard of the living (✓). More food is needed to feed all the tourists so farmers have more people to sell their crops to and this increases their profits (✓). The things built for the tourists like water parks can be good fun for local people too (✓). Tourists cause bad things as well. Tourists make lots of noise when they go out clubbing and disturb the locals (✓). Tourists drop litter on the beach which locals don't like (✓).

Comments and marks

This is a very good answer. A case study is mentioned and both advantages and disadvantages have been mentioned. The effect on both people and the environment has been mentioned, so this answer achieves the full **6 marks out of 6**.

Key words and associated terms

Ecotourism: The main aim of this type of tourism is to educate tourists on ecological issues such as conservation and the protection of wildlife and ecosystems.

Mass tourism: Refers to the global tourist industry involving hundreds of thousands of tourists visiting both developed and developing countries of the world.

National Parks: Areas that are designated as being under the control and protection of National Park Authorities but are open to the public.

Regulatory boards: Boards designed to set controls on the impact of tourism on natural environments.

Sustainable tourism: This involves monitoring the impact of tourism on society, culture and local ecology and ensuring that it has a positive rather than a negative effect on these areas.

World distribution of diseases

Key points !

* You should be able to identify the main diseases of developed and developing countries, including heart disease, cancer, asthma, AIDS, malaria, cholera, pneumonia and kwashiorkor.

* You should also be able to describe their distribution and their causes.

Figure 3.15 shows the world distribution of diseases.

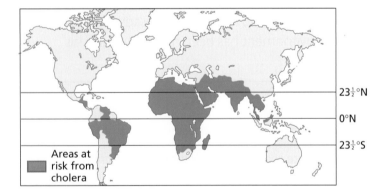

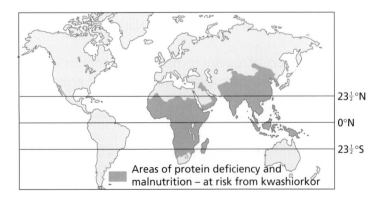

Figure 3.15 World distribution of diseases

By studying Figure 3.15 you should be able to describe certain patterns in the distribution of these diseases, which should give some clues as to their causes.

Developed world diseases: Main types and general causes

Diseases such as heart disease, cancer and asthma are more typical of developed countries.

Heart disease and cancer have similar causes, mainly related to genetic factors, environment and lifestyle.
- Genes are inherited and can be a major factor in a person's risk of suffering from heart disease such as angina, heart attack and heart failure.
- Hereditary factors are also important in some cancers and types of asthma.
- Medical research has proved that smoking is a major cause of lung cancer and heart disease.
- Lifestyle factors such as fatty diets, lack of exercise and stress can contribute to an increased risk of heart disease and cancer.
- Environmental factors such as air pollution (from sources such as traffic exhausts and industrial smoke), food additives and the use of chemical fertilisers can all affect people's chances of developing cancer, heart disease and asthma.

Developing world diseases: Main types and general causes

Diseases such as cholera, malaria and kwashiorkor occur more frequently in developing countries. General causes include:
- a lack of economic development leading to widespread poverty
- poor standards of hygiene and sanitation
- lack of clean water
- poor housing conditions
- lack of adequate food supplies, for example due to famine
- local climate conditions
- political problems
- poor standards of healthcare and medical provision
- lack of fully trained medical staff.

All of the above factors contribute to the occurrence of these diseases. The populations of developing countries are much more likely to contract and perhaps die at an early age from these diseases than those in the more developed countries of the world.

HIV/AIDS

Most scientists believe that HIV originated in west central Africa during the early twentieth century. Since its discovery, AIDS has caused nearly 35 million deaths in developed and developing countries. In 2017 it was estimated that approximately 35 million people throughout the world have contracted the disease.

HIV/AIDS is a disease of the human immune system caused by the human immunodeficiency virus (HIV). Once a person contracts HIV they may experience flu-like symptoms for a short while but most people have no symptoms at all for a long time. However, as the illness develops it interferes more and more with the immune system until the final stage of the disease – known as acquired immunodeficiency syndrome (AIDS). Sufferers are then unable to fight off infections and tumours that would not pose a serious threat to people with normal immune systems. The infections that a person with AIDS suffers, and eventually dies from, depend on what micro-organisms are present in their environment.

How is HIV spread?

The disease is spread in a number of ways, including:
- unprotected sexual contact
- exposure to infected body fluids or tissues
- from an infected mother to child during pregnancy, delivery or breastfeeding
- needle-sharing with an infected person during drug injection
- unsafe medical injections in areas of the developing world such as sub-Saharan Africa
- infected blood transfusions.

In developed countries the risk of acquiring HIV from a blood transfusion is extremely low due to improved donor selection and HIV screening of blood donations.

However, in some developing countries where incomes are low, only half of the blood used for transfusions may be appropriately screened. In these areas, it is estimated that 15 per cent of HIV infections come from the transfusion of infected blood.

There are also significant risks of infection during invasive surgery, mainly in developing countries.

It is not possible for mosquitoes or other insects to transmit the disease. There is also a misconception that AIDS can be spread through casual non-sexual contact.

What effects has AIDS had on society?

- AIDS has had a great impact on society both economically and socially.
- Since AIDS is a progressively debilitating disease, infected people eventually find themselves unable to work.
- They may become dependent on state benefits if these are available.
- Socially, many AIDS sufferers are the victims of discrimination both at work and in society in general.
- Some religious and other groups are less than sympathetic to those who are diagnosed as HIV positive, in certain cases even regarding the disease as some sort of punishment for their 'sins'.

What strategies have been used to manage the disease?

- As of 2017 there is still no effective vaccine against HIV infection, and no cure.
- Treatment consists of highly active antiretroviral drugs, which slow the progression of the disease. These drugs consist of a cocktail of at least three different types of medication. The drugs are expensive and are often unavailable in developing countries.
- Those infected with HIV may also be vaccinated against 'opportunistic' infections such as hepatitis A and B.
- Expectant mothers who are infected are also given preventative treatment to help protect their unborn child from infection.
- The medical treatment available to a sufferer is therefore related to the economic status of the country where they live. Effective treatment is much less readily available in developing countries. In some cases, particularly in developing countries, treatment is less effective due to:
  - poor access to medical care
  - inadequate social supports
  - mental illness and drug abuse
  - failure to keep to a regimen of taking medications.

Preventative programmes

Many countries have introduced programmes that aim to educate people about HIV/AIDS and how to take precautions that will limit its spread. Some of the measures used by such programmes have included:

- encouraging the use of condoms during sexual intercourse, which has reduced the risk of HIV transmission by approximately 80 per cent over the long term, mainly in developed countries
- education on the dangers of needle-sharing
- comprehensive sexual education provided at school, which has helped to decrease high-risk behaviour.

Have these strategies been effective in the control of AIDS?

- The control and prevention of the spread of the disease has been much more successful in developed world countries. This is due to the wide availability of effective medical treatment, programmes to educate the population in preventative care, and on-going medical research programmes. In developing countries the spread of AIDS is pandemic.

- The sub-Saharan region of Africa is the most badly affected.
- About 66 per cent of all new HIV cases in 2015 occurred in this area and 25.6 million are living with HIV.
- South Africa has the largest population of people with HIV of any country in the world, at 7.1 million. South and south-east Asia is the second most affected area in the world, with an estimated 5.1 million cases. Approximately 2.4 million of these cases are in India. Prevalence is lowest in western and central Europe, with 0.2 per cent of world cases.

Diseases prevalent in developed countries

Having chosen any one of heart disease, cancer or asthma, you should be able to discuss the main causes of your chosen disease.

Heart disease

Some of the general causes have been noted earlier. The main causes of heart disease include the following:

- People whose parents or grandparents suffered from heart disease have a greater chance of developing heart problems than those whose family members did not suffer from such conditions.
- Heavy smoking can raise the blood pressure and lead to the formation of blood clots, both of which can result in heart attacks.
- Diets that include high cholesterol and fat content and/or excess alcohol can lead to blocked arteries (angina), which if not treated can eventually lead on to heart failure.
- People who are overweight and fail to exercise may be more prone to heart disease.
- Stress can lead to high blood pressure, which can in turn result in heart attacks.

Cancer

- There are many different types of cancer, which affect different parts of the body.
- Smoking can result in lung cancer, which can then spread to other organs of the body. It also increases the risk of many other types of cancer.
- Other lifestyle choices, such as high consumption of alcohol, are associated with an increased risk of other cancers, such as cancer of the liver and mouth.
- Living in an environment that is badly polluted, for example near heavy industry or near high-voltage power lines, has been suspected of being a cause of some cancers such as leukaemia (cancer of the blood). However, further research is needed before a causal link can be proved.

Asthma

Asthma is when the muscles around the walls of the airways tighten so the airways become narrower. Sometimes sticky mucus and phlegm build up which can further narrow the airways. This causes the symptoms of asthma which include: coughing; wheezing; shortness of breath; tightness in the chest.

Reasons for developing asthma include:

- family history of asthma
- eczema
- allergies to things such as dust mites or irritants in the workplace
- environmental factors such as air pollution
- modern lifestyles such as changes in housing and diet (possibly even a more hygienic environment)
- smoking during pregnancy can increase the risk of a child developing asthma and children whose parents smoke are more likely to develop asthma
- viral infections.

Figure 3.16 summarises the main causes of heart disease, cancer and asthma.

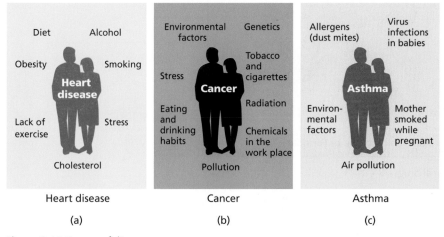

Figure 3.16 Causes of disease

Diseases prevalent in developing countries

Malaria

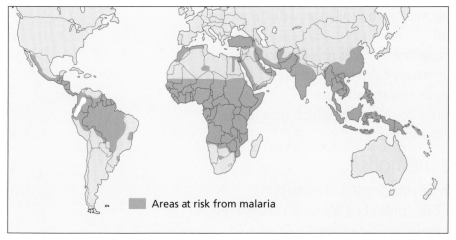

Figure 3.17 Areas at risk from malaria

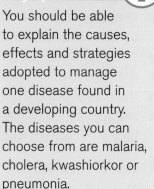

Key point

You should be able to explain the causes, effects and strategies adopted to manage one disease found in a developing country. The diseases you can choose from are malaria, cholera, kwashiorkor or pneumonia.

Malaria is caused by a microbe and spread by a vector (insect) that carries the disease, namely the female *Anopheles* mosquito. The mosquitoes pick up the disease through taking blood meals from infected people and pass it on in the next blood meal through their saliva. These mosquitoes breed in stagnant water, for example in marshlands, under certain climatic conditions. Suitable conditions are generally hot, wet climates, for example a minimum temperature of 16 °C. The disease can spread very rapidly throughout an area unless measures are taken to limit and control it.

Mosquitoes have become resistant to many insecticides and the malaria-causing microbe itself has adapted to become resistant to certain drugs that were formerly used to cure it. As yet there is no effective vaccine available to prevent infection, although great efforts are being made to produce one in various medical research facilities.

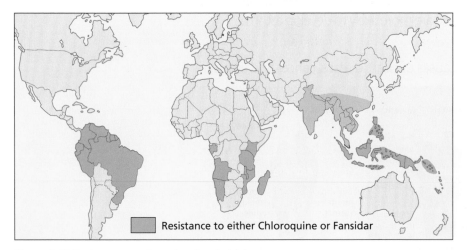

Figure 3.18 Resistance to selected anti malarial drugs

Methods used to control and manage the spread of the disease include:
- drainage of areas with stagnant water, for example swamps, and water management schemes to destroy breeding grounds of mosquitoes
- the use of insecticides such as Malathion
- the use of nets to protect people while they are sleeping
- the use of drugs to control the disease, for example quinine and derivatives of this drug, such as Chloroquine
- the use of village health centres and information/education programmes through **primary healthcare** schemes
- releasing water from dams to drown mosquito larvae (which need to come to the surface to breathe)
- using egg white sprayed onto stagnant surfaces to suffocate larvae
- the introduction of small fish in paddy fields to eat larvae
- the planting of eucalyptus trees to absorb moisture
- the application of mustard seeds into the water, which drag larvae below the water's surface and drown them.

How effective are these methods?

- These measures have met with varying degrees of success.
- No effective vaccines have yet been produced, although several test studies in (for example) China are achieving progress.

- Much depends on local populations having access to and committing to suggested precautions and medications.
- Malaria unfortunately remains a significant debilitating and lethal disease in many parts of the developing world.

Cholera

- Cholera is a form of gastroenteritis caused by bacteria in contaminated food and water.
- There are many different types but symptoms for all include: severe diarrhoea; vomiting; dehydration and leg cramps. It has a mortality rate of 30 to 50 per cent without treatment. In some types of the disease, a person can die within two hours of becoming ill.
- Cholera can spread extremely quickly in areas of poor sanitation. It is therefore mostly found in developing countries which have poor water supplies, sewage and drainage. It is rare in developed countries which have better access to a reliable, clean water supply and good drainage.

Methods used to manage and prevent outbreaks of cholera include:
- establishing and maintaining clean water supplies and hygienic waste disposal systems to prevent water supplies becoming contaminated
- education to teach people how to prevent infection – for example, by washing hands after going to the toilet, thoroughly cleaning food and cooking it properly, boiling or sterilising water before drinking it
- there are several different vaccines available which are often given to travellers to prevent them contracting the disease though most developing countries cannot afford these for everyone
- treating infected people with rehydration salts or intravenous fluids and/or antibiotics while also ensuring they don't pass the bacteria on to others.

How effective are these methods?

- In most of the developed world, good water supplies, sanitation and education has meant that outbreaks of cholera very rarely occur.
- However, many developing countries which do not have the infrastructure to provide safe water supplies are affected. It continues to be a major problem in several Asian countries as well as in Africa.
- It is particularly common where water supplies are disrupted through disasters such as earthquakes or war.

Kwashiorkor

- Kwashiorkor is a form of malnutrition which occurs when there is insufficient protein in the diet.
- It happens where there is famine or limited food supply caused by poverty, drought (or other disasters), neglect or lack of nutritional education. It is mainly found in the least developed countries in the world, especially in parts of Africa.
- Symptoms include: weight loss and poor growth; a large swollen belly; enlarged liver; changes in skin pigment; dermatitis and diarrhoea. As well as death, it can also cause permanent mental and physical disability.

Methods used to prevent and reduce the prevalence of kwashiorkor:

- ensuring that people have access to the right food; to prevent this disease, 12 per cent of calorie intake should be protein
- improving crop yields and food supply through improved irrigation and prevention of soil erosion
- producing cheap sources of protein such as soya, sunflower and maize which poorer people can afford
- storing food so supplies are maintained even in times of disaster
- education to improve farming techniques as well as nutrition.

How effective are these methods?

- In some places improved farming techniques and growing cheaper protein crops have successfully improved access to food.
- However, because the best sources of protein (dairy products, meat and fish) are too expensive for the poorest people, it remains a huge problem in some parts of the developing world.
- Larger farms in many developing countries grow profitable crops to send abroad, rather than cheaper ones which poor people in their countries could afford.
- Subsistence farmers struggle to produce enough protein in areas of the world commonly affected by drought.

Pneumonia

- Pneumonia means inflammation of the lungs caused by infection. It is common among the young or old and people with serious underlying illnesses. It is the largest single killer of children worldwide.
- Symptoms include: coughing; shallow breathing and shortness of breath; high fever, chills and sweating; diarrhoea and vomiting.
- It occurs in both the developed and developing world but is less fatal in the former due to better access to health care and antibiotics.

Methods to prevent and manage pneumonia include:

- improving nutrition and living and working conditions by reducing air pollution as well as preventing diseases such as HIV which can cause pneumonia
- education so people know the symptoms and can therefore access health care early enough for treatment to be effective
- better and cheaper access to health care, including mobile health clinics and visiting people's homes so pneumonia can be treated.

How effective are these methods?

- Several aid organisations, including UNICEF, have made tackling pneumonia a priority and they have met with some success with things such as mobile health clinics and publicity to raise awareness.
- However, for the world's poorest people, inadequate diet and bad living and working conditions mean they are still highly likely to contract pneumonia and cannot access or afford the health care necessary to treat it.

Example

World Malaria Day

25 April

#LetsCloseTheGap

Global stats on malaria		
Cases	**Incidence**	**Mortality**
212 million	21%	29%
In 2015, there were 212 million cases of malaria worldwide.	Between 2010 and 2015, there was global decrease in malaria incidence.	Between 2010 and 2015, there was decrease in global malaria mortality rates.

Figure 3.19

Look at Figure 3.19

For malaria, cholera, kwashiorkor or pneumonia describe, in detail, the effects of the disease on people.   **6 marks**

Sample answer

Malaria causes thousands of deaths every year (✓). People suffer from fever, chills and sweating (✓). Children are sick so cannot go to school and miss out on their education (✓) or they miss school because they have to look after their parents (✓). If the parents are ill they cannot go to work so the family become poor (✓).

Comments and marks

This is a good answer. It mentions both the physical effects on people as well as how their life is affected by the disease. The answer is a bit short. It needs more detail in it to achieve full marks. This answer is worth **5 marks out of 6**.

Key words and associated terms

Primary healthcare: This is a system designed to provide basic health and medical care that is cost effective and readily available to people suffering from relatively minor health complaints in developing countries. Rather than using highly trained medical staff or expensive hospitals, it relies on people who have basic medical skills and is therefore available to a larger proportion of the population.

Student materials

You will, no doubt, have a variety of classroom and online resources available to help you with your revision. Here are just a couple of the more popular and widely used ideas, should you wish to use them.

Spidergrams

Spidergrams (or spider diagrams) are a very useful study tool because they help you to organise information in a way your brain more easily absorbs. This is partly because the structure of spidergrams shows the relationships and hierarchy of ideas within a given topic. Put simply, the patterns that form in a spidergram visually show the links between ideas, and how closely connected they are.

Below is a real student example investigating the theme of prediction methods/planning for earthquakes.

Example student spidergram

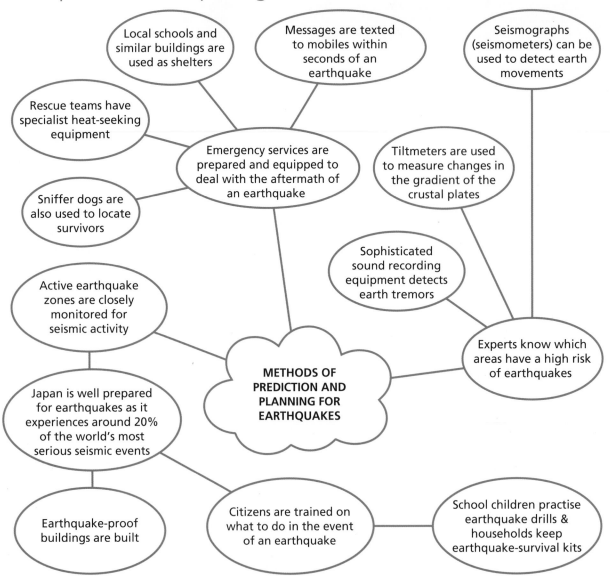

Can you identify the three key areas coming from the main theme in the central 'cloud'? Do you agree with them? Can you think of any others? How would you structure this differently?

My spidergram

Why not have a go yourself? On the next page is a blank spidergram for you to add your ideas to. We have drawn three blank circles coming from the centre as you should try to think of at least three ideas that link directly to your central topic/question. However, there is space for you to add more, should you wish.

Write your main topic or question in the central cloud shape and work outwards from there. (If you do not own this book, you are permitted to photocopy that page for your personal use.)

PP= 3 corries,

Checklists

Another useful revision tool is a progress checklist. This is a simple chart that helps you monitor the topics you've revised to date, along with how well you feel you have understood them.

'I can do' self-assessment checklist

There are many out there, but a very easy and effective idea is the 'I can do' or 'traffic-light' checklist. It works like this:

RED: means that you don't feel that you have understood this topic very well/at all and that you'll need to go back over it again at some point.

AMBER: means that while you've grasped some (or most) areas, there are still some aspects you are having problems with. You will need to revisit this to make sure you have fully understood the topic.

GREEN: means you encountered little or no difficulty with this topic and feel confident you fully understand all the key areas.

There is also a blank comments box in each row for you to add in any notes that might be useful to you about that topic.

	Red	Yellow	Green	Comment
Section1: Physical Environments				
1.1 Weather		⬭		
1.2 Physical landscapes	⬭			
1.3 Land use related to the four specified physical landscapes				
1.4 Conflicting land uses and strategies to manage them				
Section 2: Human environments				
2.1 Population		⬭		
2.2 Urban areas	⬭			
2.3 Rural areas	⬭			
Section 3: Global issues (any two from the following:)				
3.1 Climate change		⬭		
3.2 Natural regions				
3.3 Environmental hazards				
3.4 Trade and globalisation				
3.5 Tourism				
3.6 Health		⬭		

We've devised the checklist above to cover each of the chapters in this book. You might want to use this as a way to keep track of your study progress. Alternatively, you might want to devise your own checklist, either individually or with the help of your teacher. There are no rules – whatever works best for you.

Don't forget - if this is not your book, you are allowed to photocopy this page to use instead.

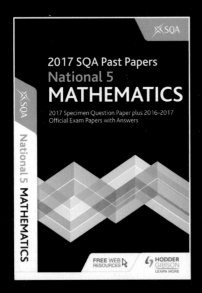
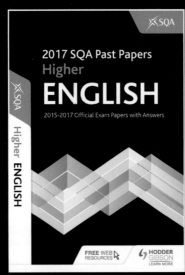
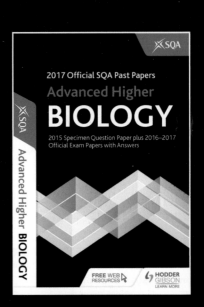